The
Brain
Is
A
Wonderful
Thing

**Dedicated to
Suesie Q
Without whom
life would have circled.**

The Brain Is A Wonderful Thing

Lee Kent Hempfling

Published By
The Enticy Institute
P.O.Box 6932
Apache Junction, Arizona
85278

CONTENTS:

Preface:

Each chapter of this book was presented as a separate paper with solicited comments and reviews possible at EnticyPress.Com. Each paper was initially in draft form.

This is a work of knowledge and thinking, not grammar and punctuation. If you find problems with grammar, sentence structure, use of commas, tense, or syntax, ask yourself first, 'was it meant to be that way?' and if not: Get over it. The topics are far more important than the presentation.

Regardless of how informative this book may be, the entire premise requires thought. It cannot be read at the speed one may be accustomed to reading.

When reading this book it will serve the reader well, to look at each and every word, one at a time, while saying the word or 'reading' the word. It even helps to read each syllable of each word.

The reason is that reading can become long-term processing quite quickly, especially for content the reader has no current association with. One will need the evaluative strength of short-term processing to engage these concepts. Long-term memory will reject them as not known.

By 'engage', I do not mean attack, or commence; rather, 'engage' means to allow the concepts to enter your brain without the automatic rejection, prompted by the lack of, ever hearing this topic, presented this way: before.

It is my hope and prayer that this book will help lives. It already has helped those who have read it.

lkh

Chapter One
Turn It Upside Down

Getting Started:

If you look to the image at the right, and take the position of being at the bottom of the image, where the large wave is, and consider that you are looking at a wave with momentum, away from you, you would be looking at the current model of physics.

We are comfortable with its perspective. We know that from our vantage point in the Universe, everything seems to be moving away, and all at the same rate, but the further away it is, the faster it appears to expand.

We see the same thing in light, as it travels through space to us; where the further away it is from the source, the longer the wavelength is.

Our science is based on finding the 'source' of the smallest parts of the largest wavelengths.

Yes, as we are looking for smaller and smaller particles we are looking for smaller and smaller turtles.

In looking for smaller, and concentrating on the particle, we have ignored the wave and what 'smaller' actually is.

We have been seeking 'quanta' and the causes of 'quanta' as if they were what their results are.

What we see as a wave, and visualize as a moving 'thing', is, as any grade school child will tell you, a ripple in the pond, not the pond.

We know that waves are disturbances in a medium but we think that only applies to Newtonian Physics, things we can observe with the 'naked' eye.

We study the chemical make up of matter, rocks and historical artifacts by defining the wavelengths of the compounds that make up that 'matter'.

When doing that, we are seeing frequencies of compounds that are without amplitude.

Frequencies with amplitude tend to appear to be 'alive'.

'Alive' can be electromagnetic and cause friction and destruction and depletion or it can be 'Alive' while being: simply existing.

'Simply existing' is a good term to refer to a non-zero wavelength with no amplitude.

Our 'perspective' affords us a view that we are able to calculate the existence of a non-zero value energy 'charge' that is not the energy 'charge' of 'this' perspective.

Call it 'vacuum energy' or 'quintessence' (but, don't keep that word: it completes the current theory of everything qualification model as a direct clone of earth, wind, fire, water and 'quintessence', held up by an elephant standing on turtles.

The image to the right is a closer rendition of what the evidence points to in cosmology. Taking the same perspective, by standing at the bottom of the graphic and looking 'out', we can see that our visible Universe is made up of two charge 'states'.

We are familiar with one of them and just recently became 'aware' of the other one.

The wavelength we see at the 'end' of our 'visible' Universe appears to be far wider than logic would assume.

We have contemplated this 'exponential' process and have been unable to explain it.
Taking the exact same process, and looking at it much closer, can explain it.

As we 'measure' away, on one of the 'red' electromagnetic charge 'states', and observe its frequencies in various applications and methods, it is very logical to completely miss the non-zero value of the other charge 'state'. It has what appears to us to be 'no' wavelength

Not 'seeing' the other charge 'state' we, as humans have eluded to its existence through religion and mysticism and a 'higher' power. Yet, we have not known it scientifically for what it really is, as we have never 'observed' it in a Newtonian condition, regardless if we find evidence for something 'weird' in the rather interesting causes of history's settling on the particle model for quantum mechanics.

Looking at the interaction of the two charge 'states' from a 'side' view, the model can represent the interaction of any two 'things', whether they are people and relationships, or atoms, or particles, or anything else there is, that interacts.

If we were to show this interaction in a more 'human' manner, one less 'electronic' in appearance it would look like this:

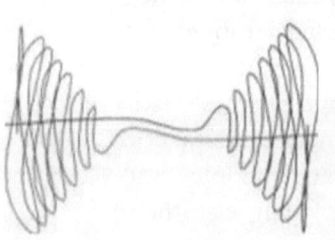

Closer together, we can see how the waves of both start to reach more wavelength intensity. If the red in this image represented one person and blue represented another person and they were that closer in a relationship or in a common interest a point is reached where the existence of both represent a sort-of 'harmony' we might refer to as 'friends' or the 'relationship', where the affects of that interaction are what the other 'party' perceives of it.

If we manage to get 'really' close in a relationship the waves (the closer to the same frequency the better) reach closer to 'source' and they overlap.

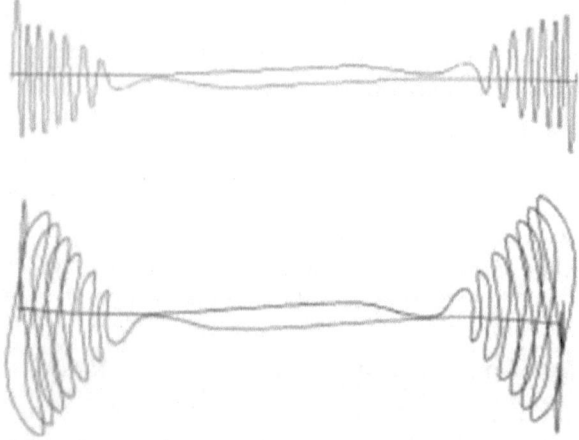

That is true love. It is also a chemical bonding, an attractive force of gravity and the opposite of the repulsive force of different frequencies. Like frequencies attract, opposing frequencies, repel.

Closer together, we can see how the waves of both start to reach more wavelength intensity. If the red in this image represented one person and blue represented another person and they were that closer in a relationship or in a common interest a point is reached where the existence of both represent a sort-of 'harmony' we might refer to as 'friends' or the 'relationship', where the affects of that interaction are what the other 'party' perceives of it.

If we manage to get 'really' close in a relationship the waves (the closer to the same frequency the better) reach closer to 'source' and they overlap. That is true love. It is also a chemical bonding, an attractive force of gravity and the opposite of the repulsive force of different frequencies. Like frequencies attract, opposing frequencies, repel.

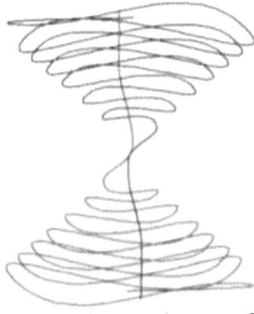

We have been looking at the Universe from our perspective. Our perspective represents a small minority of the Universe, we now know exists. In order to comprehend the perspective of interactive bodies we must turn it upside down:

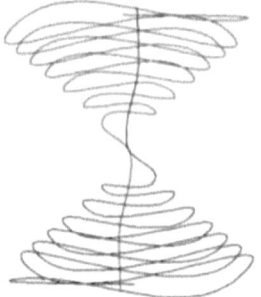

From this perspective, we can see, that regardless of perspective, it works the same way, but our perspective has prohibited our observation of the potential of a true and complete and most simple relationship as until now, we have not bothered to look for the non-zero value of 'dark energy', seeking in turn to 'find' the 'meaning' and 'source' of what we think is not as simple as an opposing charge 'state'.

The simple relationship is one of merging via influence.

$$\frac{a - a'}{2} + a'$$

It is, a simple 'average' with one difference. It has order.

In a 'simple average' the order of the part and the count of the parts are not relevant. In the Neutronics Equation the order makes the equation an average with logic.

The logic is order. Whichever 'part' is causing the equation to exist is the 'actuator' represented by **a'** while the 'perceiver', represented by **a** is receiving the 'actuator'.

Working a reverse calculation, turning it upside down, allows the opposing leverage of the relationship.

The perspective from which the measurement is made determines the order of the observation, but not the order of the equation.

Only by, 'turning it upside down', can we observe the 'concept' of the 'actuator', or the 'other person' or the manner in which we view our science.

By applying this equation, searching for a model from which to base its understanding has resulted in an awareness of the exact same equation and 'system' at work in everything there is, including the brain.

If need be, read this chapter again. Stay short-term and focused on the words as you read them.

Chapter Two
The Brain Is A Wonderful Thing

Defining, Demonstrating and Detailing Brain Function And Consciousness.

"There is much to like in this article."
Joseph Naimo, Ph.D. Murdoch University Western Australia

The brain is a wonderful thing. It gives us the ability to see, hear, sense pressure and temperature as well as taste flavors and appreciate smells, remember, recall, evaluate and deceive ourselves that *we* are doing those things.

Since it is the single source of processing in the body that allows us to experience these wonderful sensations and to be aware of the experience it is also the cause of everything we know, everything we believe and everything we do and say.

This chapter will detail how the brain processes the information it has, what that information is and in so doing clear up a slight misunderstanding of Uncertainty and present a scaled down version of a human brain working on the Internet.

The Brain Is A Wonderful Thing

Perhaps above all others the human brain's most astonishing achievement is the ability to give its host body the ability to know.

Evolution Of The Brain

Input Process Output	In Out	
Input Compare Output	In Small Long-Term Out	
Input Compare Output	In Taller Long-Term Output	
Input Compare Output	In Taller Long-Term Out	
Input Compare Process Combined Output	In Taller Long-Term (In-Reference) Combined Output	
Input Compare Compare Combined Output	In Tall Long-Term (In-Reference) Short Non-Looped Short-Term Combined Output	
Input Compare Compare Loop Combined Output	In Tall Long-Term (In-Reference) Looped Short-Term Combined Output	

To know, is not just to collect data. A computer does not know it has data even though it may report that it is missing some. Any brain other than human does not know.

After the initial non-zero wave gained amplitude the brain's evolution took it through seven distinct progression changes. Those changes added layers and processes to the brain with the resulting additional abilities attributed to them. Later in this piece you will see how the Mattel® game Uno®; will show you how this works.

From the simple input (process with the base amplitude) to output the brain has evolved by first adding memory (which later becomes long-term memory).

This occurs through the amplitudes processed in memory pathway neurons. At the end of the memory path, if the amplitude is large enough the output will cause another neuron to grow at that point to process it.

We observe this action today with regeneration.

This growth continued from stage 2 evolution through stage 5 where increases through evolution of brain 'power' are seen as depth of memory (later, long-term memory) and recall ability and the balances between firing rates and orchestration of blending patterns through level processing ratios.

Each stage developed over time before reaching the next stage. This graphic depicts full state condition of each stage.

In stage 5 a second process level was introduced acting upon a further reduction of clock rate causing a faster processing rate than long-term. Two outputs then worked together to provide a greater degree of control over motion but without short-term memory to retain the result of the second level of processing very little additional control was provided..

In stage 6 short-term memory grew from the outputs of short-term second level processing (same generation e.g. regeneration as above) causing a retention of short term processed amplitudes and building up a second level process that did have application to motion control. Short term then was able to retain and return loops of recall but was not a closed system to force recall of recalled recent memory.

In stage 7 short-term memory grew to become the largest part of the physical make up of the brain as much longer pathways were needed to process and retain higher amplitude impulses when processing was much faster than long-term. Long-term has remained rather stable since stage 5 .

In stage 7 the major change provided by evolution is the connection of the end of the short-term pathway to the beginning of the short-term pathway through a feedback return loop that was both long enough in depth of retention to allow a set up of redundancy as default and the change in the input to long-term memory.

It is important to note that terms used in this piece may (and have) caused some concern about the meaning and utilization thereof.

Two of the confusing aspects of dealing with brain function and the dynamic system are:

1: understanding the degree of effect one level of processing has on subsequent levels of processing. This paper will detail the ratio enhancement aspect below but may not suffice to impart a grasp of the concept of intelligence itself.

2: The notion of 'awareness' or 'knowing' is described herein but may not provide reference to readily known comparisons and therefore not suffice to impart a grasp of the concept of intelligence itself. To that end these graphics depict the levels of human brain processing and the importance of what a loop does to create the 'awareness'.

Exponential Processing

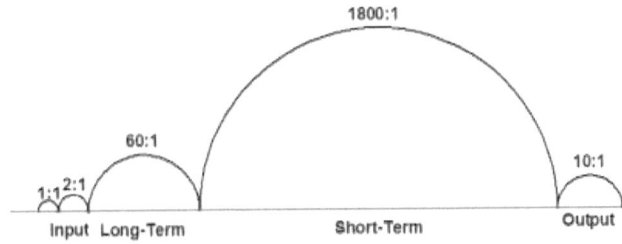

1:1 Your computer performs 1:1.
2:1 Input samples twice per second
60:1 Long-Term memory compares to input 30:1 per half second
1800:1 Short-Term memory compares to Long-Term memory 900:1 per half second
10:1 Both Long and Short-Term memory outputs are combined in reduction to motion

Loop Awareness

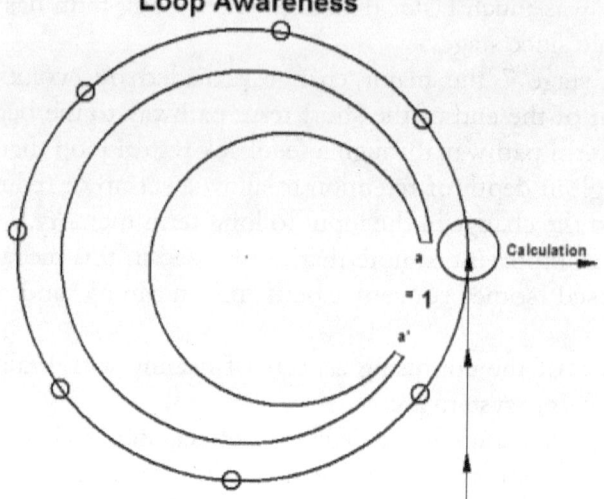

Understanding the degree of effect one level of processing has on subsequent levels of processing requires understanding exponential processing:

In the upcoming white-paper 'On The States Of Charge, Gravity and The Observable Exponential Universe' by the author the concept of exponential events is detailed describing the observed multiplicative and additive inverse conditions which requires completion of equality or the reached state of additive or multiplicative identity. Chapter One of this book begins that discussion.

A dynamic system such as the brain has degrees of variance between 1 and 0 identities and does not become 1 or 0 until equality is reached in amplitudes.

It is that degree of equality difference represented in the dynamic system of the brain as amplitude variations that reaches both multiplicative and additive inverse conditions only when the pathway is closed and looped.

Partial loops occur throughout both long and short-term memories and account for recall ability and habits, depression and even the existence of mentally caused physical maladies represented by Stigmata but the reaching of a totality of 1 or 0 of the loop is determined by the amplitudes and that requires depth or 'height' of pathway processing to accomplish.

A seriously suffering depressed person will have a loop close in length to the total awareness loop in short-term memory that is based on a supportive long-term memory return loop fed by that short-term loop.

Humans are humans because we have a 'hard-wired' loop at the fullest extent of the short-term memory which forces all memories in short-term to loop back to the origin of processing not allowed in the extended non closed looped long-term memory.

Imagine a rolling ball that gets smaller and smaller as it rolls in a circle.

Long-Term Memory   Short-Term New   Short-term Aware

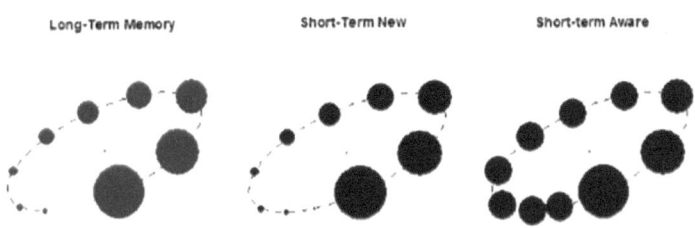

That would be long-term memory.

Imagine a ball that got smaller as it rolled but each time it reached the end of the circle it bumped into the next ball and that ball grew by the mean sum of both balls.

Then imagine the importance of any ball never being allowed out of the system. (Short-term's closed loop.)

Over time (4-7 years normally) the size of each ball will be half the size of the ball it hits when it completes the circle which will be a closed loop self-aware brain.

That loop will happen in the strongest pathway type.

Short-term is provided for all major inputs but is most longest (by massive increase) in aural and visual pathways.

Whichever of those two major pathways reaches 1 results in its becoming the dominate form of thinking employed by the subject.

So there are aural and visual thinkers and they do not see the world in the same way.

Uno® is a trademark of Mattel® Inc.

The easiest way to visualize the difference between long and short term memories and how the two work together in humans is to play a modified version of Uno®, by Mattel®. Uno® comes out of the box in four separate game card packs.

It is possible to use an older version of the game as long as the order of the cards matches the order of each other pack of cards. Separate the cards by color. Place them in the same order.

Two people play the game by each having two color decks. Stack them on top of each other so each player has a two deck pile.

To play this version of Uno® place the two deck pile in front of you and draw the top seven cards as your hand. Make the next card in your pile your face up card to start the discard pile.

Each player has their own discard pile and their own draw pile. Play begins with either player going first. Play alternates in turn.

Play occurs from the opponents discard pile.

If player A goes first and has a yellow 3 then the opposing player goes next based on that yellow 3.

It is just like playing regular Uno® except you have your own cards, you play your own hand, and the discard pile of the last play is the play pile.

Place your discards on your own discard pile.

As the game progresses do NOT shuffle your cards (either player): simply turn the discard pile over and start from there.

What you are doing is exercising your short-term processing to decide which card to play based on the rules of the game (which still apply) and the result of what you choose are placed in the discard pile.

The discard pile remains in order as return long-term memory remains in order. The discard pile becomes the new card pile and the results of your choices are placed in that long-term memory.

You have duplicated the process of long-term memory represented by the turning discard pile and have used short-term memory to play the game.

If you were a dog you would play the game only as the cards were dealt without choice.

The order in which those cards would return to be played again would be very similar to the order in which they were first dealt. All life is dealt a deck of events based in environment as well as genetics and the result over time can be very 'instinct' looking.

It would be impossible for any person to comprehend the lack of short-term processing as to be without it is not to know you are without it.

This Uno® game demonstration is able to show you how long-term memory works and as a result of short-term processing will come back to haunt you if you make the wrong decisions or violate the rules of the game along the way.

At first the game will appear to be absurd since all colors are staying together and numbers are close to their order.

After numerous hands it will be noticed that the order of the overturning discard pile of each player begins to appear different and with more plays will begin to set up a pattern of play of that player and in turn be the style the player plays in.

Personality is a development of short-term decisions lodged in long term facing unfamiliar circumstance resorting to style.

Of course this version of the game is not intended to be a hit at your next party.

If you shuffle each player's two deck pile (four players can do the game with one deck each) before starting the long-term memory game it would be indicative of the style and method of shuffle and the specificity of the process which would compare to genetic make up as well as environment.

Unless there is a genetic error causing brain miss-wiring or clock malfunction the architecturally derived dynamic system is the same.

The other difference between humans and the rest of living brains is the change of location of the processing feed to long-term memory.

All brains have internal outputs.

An internal output is a signal sent within the brain for additional processing versus a signal sent for action outside of the brain for motion.

In stage 2 through 6 of brain evolution that internal output to long-term memory comes from the process of comparison of amplitudes of long-term to input and makes long-term a result of previous long-term and new input.

In humans the feed to long-term memory is a result of short-term processing which allows humans to retain a sense of self in long-term so you are able to recall your awareness.

Exponential Processing occurs in natural systems but is not readily observable unless at either long distances or long wavelengths or both (as in cosmology). Yet it is evident in the brain as part of the dynamic system.

This occurs by sending data from one level of processing operating within a pulsed frequency to a 'next' level of processing operating in a second level of pulsed frequency. It causes samples to be divided by the equation method shown below.

John Ellis of CERN writing in "Vacuum Energy: A Naturalness Challenge" stated, "...clarifications of the magnitude of the vacuum energy and its equation of state are of crucial importance for fundamental physics, as well as for cosmology." 'On The States Of Charge, Gravity and The Observable Exponential Universe' deals with that equation of state and will be available for pre-publication reviews soon.

Ellis also said in the same piece, "Established theories of particle physics make contributions to the cosmological vacuum energy that are many orders of magnitude greater than its possible value.

The latter can only be calculated within a complete quantum theory of gravity, to which the vacuum energy poses a naturalness challenge orders of magnitude more acute than the hierarchy of mass scales in particle physics.

No convincing mechanism for canceling the vacuum energy has been formed in string theory, the only persuasive candidate for a quantum theory of gravity. Perhaps string theorists have been barking up the wrong tree, and should shift their attention to calculating the small non-zero value suggested by observation."

The non-zero value or condition (as it essentially represents the minimum required presence of an entity or state to cause the existence of the opposite of the existence of the opposite of that entity or state) is hoped to shed light (sorry) on cosmology's quest for understanding the Universe.

It also has implications for the study of the brain which will be detailed a bit more later in this piece and completely in the referenced paper nearing completion.

Reference is made in this book to the use of 'dark energy' the presentation thereof having raised some interest and concern from pre-publication reviews.

It is understood that unless a person has a background, knowledge or interest in electronics versus a theory of energy the importance of the schematic provided below will not be readily comprehended.

The point of the schematic is to provide a replicate-able regime whereby the presence of a viable 'vacuum ' or 'dark' energy exists and is more than the presence of a non-zero event or state.

Exponential processing utilizes the amplitudes of the inverse charge state to form calculations of amplitude executing at a pulsed rate then expands that calculation from each pulse rate out to faster pulsed rates based on the single initial rate.

The exponential growth of both speed and 'power' and the reduction of base clock pulse rates to accomplish it are most of this paper's discussion topic although the system's architecturally derived process is also discussed.

One major problem to be solved in brain research is the concept of cognition. As defined by Webster: cognition is "the act or process of knowing including both awareness and judgment".

The opposite of cognition would be to not be aware but the error in the dictionary's definition of cognition is the inclusion of judgment.

Judgment, "the process of forming an opinion or evaluation by discerning and comparing" is a process of thinking and is present in every brain. The degree to which judgment is present and the manner in which it is accomplished has nothing to do with the simplicity of awareness, It is far more complicated.

Awareness is defined as "having or showing realization, perception, or knowledge" which itself is a major error.

Showing or displaying a trait does not mean the trait is present it just means the observable outcome of the trait is able to be accomplished by more than one means.

Realization, perception and knowledge are all parts of the more complicated process of thinking provided by architecture and even though all results of brain function are a result of that architecture the use of it has nothing to do with the structure of it other than the structure results in it.

Can you imagine not being aware yet still being conscious?

Conscious is defined as "perceiving, apprehending, or noticing with a degree of controlled thought or observation" and is completely wrong.

It is why terms are interchangeable.

A person can be said to be unconscious when knocked out but sleeping when just not being awake. A person can be said to be conscious after the anesthesia wears off.

Those things apply to both human and other forms of brains yet no other form of brain is capable of being aware and because of the architecture humans are capable of controlling the realization, perception, knowledge, thought and observation.

So what then does it mean to be without awareness yet still awake, lucid and functioning?

Just think of your family pet: cat, dog, snake, fungus collection (no, forget the fungus collection).

Auggie is a sweet and wonderful blue tick hound that happens to live with my family.

Auggie is smart as are most hounds but since she is treated differently than other hounds her responses to suggestions are different.

Auggie is a dog. Tooter is cat. They both have brains and both use them to the utmost of their potential because it is wired that way. They are different in that the feline species is primarily a visual thinker while the canine species is primarily an aural thinker.

Of all the senses: vision and sound are the two primary senses and as such are the ones given the longest depth of memory while their output is to things very important to the body.

It is the same in all brains.

Input of visual stimuli, the processing of that stimuli past previous similar stimuli (compared memory), outputs to the muscles of the body resulting in motion.

Young aural thinkers tend to be clumsy and unable to take quick control over movement.

Input of aural stimuli, the processing of that stimuli past previous similar stimuli (compared memory), also outputs to motion but to the specific and detailed areas of the mouth: tongue lips, facial features: movement with specific purpose other than traversing distances or avoiding injury.

Together in a healthy brain, the visual and aural processing of the brain work in harmony to create a being that interacts with its environment and can to some degree control that environment.

The processing taking place in all brains starts with input.

What happens to the receptor sets the stage for what happens throughout the brain using that receptor.

Common folklore (common sense is examined in chapter twelve) would have one believe that sound enters the brain as a musical note or a voice or word and somewhere in that brain the musical note or voice or word is stored as a musical note, voice or word.

In reality, the receptor is not inputting anything.

The sound wave caused by the source of what is being heard is affecting the receptor set to respond at its frequency and that response controls the passing of the brains dynamic system through the receptor into processing thereby allowing just about any form of receptor to process the same brain dynamic system.

It does it by interrupting the flow of brain dynamic signal set at the specific frequency of that brain by regulating the amplitude of each wave pulse or wavelength.

What is processed is a specific amplitude (amplitude is what is controlled by the receptors) for that pathway of processing which after being processed outputs to a specific muscle or muscle group.

You are not aware of that process and it would seem a bit convoluted to consider that somehow vision is making your movements for you without your having to actually will it to happen.

The reason you are not aware of that process is that you are not aware of any process going on in your brain. You are only aware of the existence of a process.

Cognition is the awareness of the existence of a process.

It is caused by a feedback loop pathway, more accurately termed a return pathway that literally connects the end of the short-term memory of humans with the beginning of the short-term memory.

That complete circle of processing is what makes humans, human.

Auggie and Tooter do not have short term memory so they are not aware of a process which means they are not aware that they are alive, or aware that they are doing something or aware that they are at all a thing different from any other thing.

Non-human brains (minus the Quadra pedals who have varying degrees of short-term memory but lack the circle of total return) are working in long-term memory.

Long-term memory can be summed up as mostly reactionary with degrees of pro-action based in the amount of enhancement brain processing uses to evaluate it.

Can you imagine not being aware yet still being alive, still thinking the same way as other humans but just not having the awareness of the process?

Young human children are just that.

Even though the architecture provides a return loop at the end of the short-term memory in humans it takes a long time to build up memory values within the long-term memory based on the depth of that loop.

All memory returns for processing and recall. It happens just like a side-by-side divided highway with interconnecting onramps allowing a U-turn every so many yards.

Near the end of long-term memory the last U-turn onramp happens before the memory pathway is finished. That means very small amplitudes being processed in neurons, sent along axons and converted to chemical transmitters to bridge the gap of the synapse (the world's first biological diode) are falling off the end of the road and simply ceasing to exist.

For a knowledge of the awareness of self (awareness of the existence of a process) to be set up it has to find its way all the way down the long-term pathways and set up its own loop.

At first that short-term loop is working at the top levels of long-term memory. It has very little depth and is therefore more a product of short-term control therefore little control at all.

After a few years of awareness of self (awareness of the existence of a process) the loop falls deeper and deeper into long-term memory until the sense of being a unique being becomes known by both short and long term memory pathways.

While that process is taking place the child will sense some degree of difference and may attribute it to an imaginary friend or a specific blanket or a specific person.

When awareness does happen the child starts to try to fit in to environment and can only do so by testing the limits.

But can you imagine not being aware of self while still being conscious?

You are looking at a machine that is getting its signals from a machine that does exactly that.

Yes, under the definition of 'conscious' offered by Webster: "perceiving, apprehending, or noticing with a degree of controlled thought or observation" the computer you are using is conscious.

It perceives as it receives data. It apprehends as it retains data.

It notices as it acts on data.

What it does not do is pro-act before data is perceived as long as the first data was not equal to coma.

What it does not do is retain data that has been compared to all previous data and compares it to all incoming apprehended data.

It also does not know that it notices any act of its own process.

The architectural result of being in a loop that exists once as a nearly flat line frequency single wavelength the result of her mother's and father's frequencies with the merged amplitude created by the merging of them during conception and continues as that frequency wavelength turns into billions of wavelengths and then slowly converges back to one wavelength with amplitude remaining as the sum of all merged wavelets.

Luckily Auggie is still with us and going strong even though she is a tad bit overweight and prefers rugs to grass.

A computer can be programmed to do exactly the same process. As slow as real brain processing is the emulation of the enhanced levels of processing from an average twice a second for visual and aural input receptors to 20 to 30 to one increases in long-term memory processing for a very smart puppy will take quite a great deal of computer processing raw power but luckily directly proportional to the height of the long-term memory pathway stack and return pathway.

Output to motion is reduced to 5 to 12 times per second for each pathway, humans typically at 10 per second, per pathway.

The more pathways processed in a computer the more memory is needed to do the depth of field processing. It takes up a great deal of room and processing RAM to accommodate a brain near the evolutionary path level of an amphibian.

While the description of what a computer does not do could be construed to be a discussion, which included the short-term model of the looped memory, it was not.

The same concept exists in long-term memory but both at a much slower speed and with only a single output to motion.

The result of adding a short-term memory with a complete loop is to create a literal closed system. It has inputs from the computation of external stimuli and long-term memory with output to motion and long term memory the human brain is a complete system.

A top level closed system fed by a rather short long-term memory that is made up of the top level closed system's computations and dimensional, as short-term memory is, with continuation of computation all the way down the line until the signal's amplitude is so low it can no longer trigger the body to build another inch of roadway let alone a U-turn onramp.

The same applies to humans.

The major difference with humans is that humans can take in a pretty good deal of totally unrelated almost bad to devastating incidents, combine them into a general assumptive concept and let that concept rule their lives regardless as to what the reality of the current moment may be.

The solution to repair that depression caused by long-term looped supported results of external or internal stimuli is to increase the degree of control exercised by the short-term memory function and cause the long-term memory loop to first begin to break up then start to nearly completely shatter when each new realization takes place of a correct interaction and assumption.

Regression of long-term loop disbursement techniques are less and less intense and fewer and fewer in number but they will continue to bother even the most attentive short-term thinker as it is nearly impossible to trace every connection from a root cause.

Everything has a beginning somewhere. This piece, so far has spoken of the beginning of life, the beginning of awareness of life and the beginning of the understanding of the brain and how it affects future events.

It would logically be assumptive to discuss the possibility that even though you have had a few moments of uneasiness so far in reading this piece the possibility that you have had great distain and simultaneous amusement at its assertions is in direct proportion to the level of your training in brain function, epistemology, neurology, computer science, artificial-intelligence, psychology, psychiatry or religion or social services.

It is possible to make a person appear smarter.

It is possible to make a person do things that only a smarter person would be expected to do.

It is possible to control the brain of most living human beings by repeating the same thing in an unobtrusive manner in which not to cause direct connections to the point but where each connection made supports the logical conclusion of that point.

Those living human beings who fail to be controlled by repetition of subtlety are those using more of their available short-term intellect than the rest.

It is also possible to be so encased in a regime of thought that similar input stimuli representing a concept already understood in memory mixed with an input stimuli that contradicts that memory which have no potential of making a connection with the point already contained in memory as that which was previously in memory which is nothing at all like the new memory.

It would be logical to toss it out with the trash.

"Logical" has everything to do with what has been inputted the most.

It has nothing to do with new information that does not agree with past memory. Such new information is foreign and therefore rejected as not unacceptable. But what is really causing your objection to new information?

The process of brain function starts with input that is sensor specific but using the same dynamic system. Each input to a sensor causes that pathway to be planted with the corresponding dynamic system value which stands alone upon input as a single pulsed amplitude wavelet.

That wavelet is presented to a series of long-term memory wavelets where each passing long-term wavelet is compared to the input amplitude value.

If this comparison process were 1:1 long term memory would be reflective only of new input. Luckily each process of long-term past input results in a series of new combined wavelets that serves to either support previous amplitudes or reject previous amplitudes or all degrees in between.

The ratio enhancement from 1:1 to up to 30+:1 of long term memory means each input amplitude pulsed is compared to a far greater amount of long-term memory. This increase in ratio processing sets up one of the most universally experienced internal results of brain function. Time.

As memory works its way down the stack for that pathway each movement from one neuron to the next in the pathway is a result of a computation of amplitudes compared to the base amplitude. One output continues the depth of field processing while every so often a second output sends data back to be compared with input in a steady stream that skips a few returns so as to make the return line less ratio than the feed line.

The more U-Turns available the more clarity of recall is possible. Those with 'photographic' memory have far more returns to process than normal brains. Those with fewer returns to process have degrees of the inability to recall. When that happens in an Alzheimer's patient it starts at the top loop of short-term and slowly builds up the lack of awareness in long-term that is not current but rather past. As the disease progresses into long-term memory it cuts off returns near the top of the stack and deeper and deeper until the returns being fed to processing are from a time long ago and do not include any short-term process.

As memory works down it pathway like a string with knots each value is compared to the base amplitude making the value lower and lower the deeper it falls in memory. In recalling those amplitude values the result is a sense of knowledge that older memories are further away than newer memories and the result is the sense of a passing of time.

All brains have this memory process and all brains have a sense of time it is only the human brain that has a sense of 'now' from which to compare it to and therefore a sense of the awareness of a universally accepted concept.

That does not mean the universally accepted concept of the passing of time is a real thing. In fact it is a good thing as without reference of when a memory was inputted to the process no order of progression would be possible therefore no ability to return values that are of lower amplitude therefore no ability to increase those values of amplitude and no ability to learn.

Here is an example of the input process using arbitrary numbers to represent amplitudes:

Looping happens throughout the memory pathways of all brains. Undoubtedly you have been aware of another human being who had a 'rut' she was stuck in or a habit he had. And it would most probably be quite easy to detect a loop and misdirect the result away from bothersome by rubbing your cat's neck and back softly but in a near perfect rhythm quickly and then slowly slowing down to bring her insistent dry coughing to a stop. If it is a true hairball it will come out shortly thereafter.

The longer a dog is aware of another different thing the more the dog will be both accustomed to that different thing but will also have every incident observed and heard by that dog regarding that other thing working and blending with each interaction of the other thing previously to form what would either be a bad condition resulting in submissive fear to outright dangerous or a good condition resulting in a degree from tolerance to totally devoted shadow.

Here is an example of the input process using arbitrary numbers to represent amplitudes:

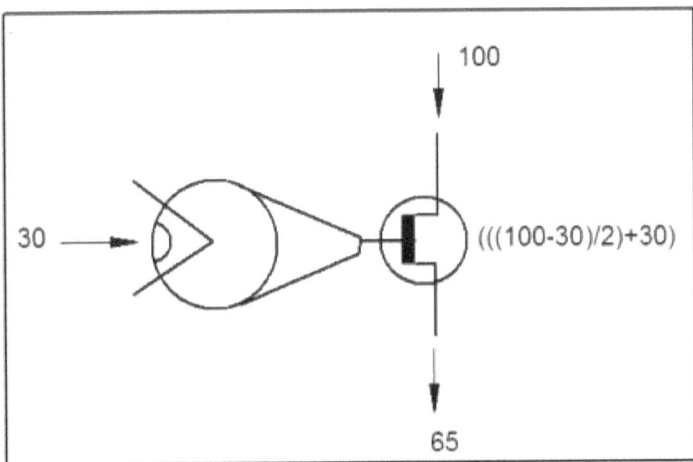

The result of the input receptor is to control the amplitude of the pulsed frequency.

That result (65) is passed by the ratio enhanced flowing long-term memory recall pathway in this fashion:

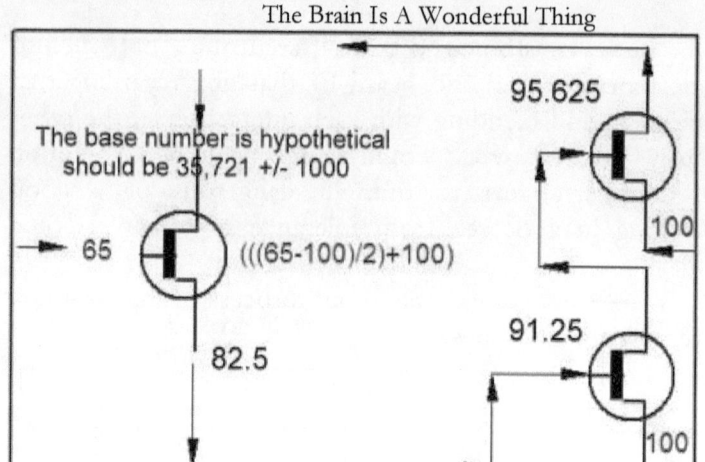

The base number is hypothetical
should be 35,721 +/- 1000

95.625

65 — (((65-100)/2)+100)

100

82.5

91.25

100

Of course there are many memory cells in a string before a U-turn is provided but this sequence gives a good indication of the process:

After each set of long-term memories is passed past the input in computation of amplitudes the next set is computed.

Notice the upside down computation process of amplitude:

Using the base amplitude of the system's "health" the receptor applies a degree of permission (versus the electronic degree of resistance) to that base and outputs the mean sum of the two amplitudes.

This is important because of the upside down nature of memory processing.

With each increase in input permission level each memory within that receptor's pulsed ratio enhanced wavelet set will decrease the memory amplitude in reference to the base amplitude.

That means a severe input exceeding the base amplitude will cause a 'shock' condition to exist in memory wavelets that being at or above the base amplitude will cause outputs to be nearly or equal to null.

It is the reason you feel pain but do not have a direct recollection of pain. Many people are able to think in short-term memory of a strawberry and thereafter retaining that concentration long enough to cause a true loop to exist in short-term regarding the topic strawberry can actually get a sense of the taste and texture of a strawberry as well as 'see' one or more.

That is recalling memory wavelets that are far below the level of null.

Since the pathways of taste, smell, temperature and pressure are stored in much shorter pathways therefore with far fewer returns the 'being' believing itself to be alive and aware of its own existence has a sense of experiencing those smells or tastes or textures as either familiar or not familiar.

It all happens within the closed system of the brain. Humans add the second closed system within the closed system of the brain and the result is knowing that one knows.

That is the binding as well as every advancement known to the planet other than evolutionary progression which itself is a product of.

Like everything else in the universe the brain is like a model of a seesaw.

Vision on one side, aural on the other, each outputting to specific motion actuators and the observed result is a new discovery detailing the "left side of the brain alone appears to take responsibility for decoding the literal meaning of emotional messages. But it seems that the brain's right hemisphere plays a role in assessing the tone in which the message is delivered - a concept known technically as prosody." [1]

"The findings are based on measuring how fast blood flows to the tissues of the brain. A greater velocity implies more activity in that area of the brain because brain cells, when active, require an increased supply of oxygen and glucose, both of which are carried in the blood. A team from Ghent University in Belgium used a technique called transcranial doppler ultrasonography to measure blood flow velocity in the brain's left and right middle cerebral arteries." [1]

"The researchers found that when participants were asked to focus on the meaning of what was said blood flow velocity went up significantly on the left side of the brain. But when attention was shifted to how to how it was said velocity also went up markedly on the right side of the brain. However, it did not go down on the left - suggesting that both sides of the brain play a role in helping to label the emotions." [1]

Blood flow is traveling to the areas working the hardest. That means the other areas not working but also involved at working at normal levels. The blood flow did not switch sides when the other side needed more blood the other side needed more blood to make sense out of the visual interpretations being developed on the other side.

The deduction is quoted as "Our findings suggest patients with right hemispheric lesions could experience more problems in understanding the emotional prosody of a spoken message and certainly may have difficulties in correctly discriminating deliberate discrepancies in content and prosody that convey more subtle forms of emotional expression in speech." [1]

Of course aural processing would be reduced with damage to that area which would result in the logic and conceptual meaning processed by aural pathways no longer properly regulating the action outputs of visual.

Add that to the degree of application the subject is using of the short-term memory process and the additional loss of aural processing control would allow degrees of direct long-term motion control to exit suffering from a degree of reduced short-term control.

Most humans go through their lives using very little short-term processing.

Outputs of non-human brains (minus Quadra pedals) are directly long-term based which results in degrees of good and bad. The long-term output of humans also results in good and bad reactions but is (or should be) controlled and regulated by short term evaluation of long-term existing memory.

Humans consider them emotions since they seem to come from somewhere other than the thinking part of the brain while all of those non-human brains including Quadra pedals are nothing but emotional.

We are aware of a thing called emotions and know it is something that happens and we pay great sums of money to people who think they know what emotions are to make them all better when the human brain has a built in anti-emotion engine.

Short-term memory in a closed loop system can work against the brain by processing the same long-term memories over and over again setting up long-term loops of distress and therefore depression or it can be used by the brain that has it to take control of those long-term memory wavelets and knock down the loops of depression.

Whether to do that or not is what anything called 'free will' would have to be.

The potential of the brain working in a dynamic system has raised enough interest to cause a massive study at Duke University.

In their opening announcement press release they said: "This research will provide us with a powerful new set of experimental tools and techniques to answer the question of how millions of brain cells come together to generate a particular behavior," ... "Traditionally, the neurosciences have taken a reductionist approach, with investigators trying to understand individual neurons, molecules and genes. We are trying to understand the brain's function as a dynamic system."

The August 15th 2002 press release cited "Brain-machine interface technologies developed under a $26 million contract to Duke University sponsored by the Defense Advanced Research Projects Agency (DARPA)".

The contract is part of DARPA's Brain-Machine Interfaces Program (www.darpa.mil/dso/thrust/sp/bmi.htm), which seeks to develop new technologies for augmenting human performance by accessing the brain in real time and integrating the information into external devices."

The Brain-Machine Interfaces Program is actually contained at http://www.darpa.mil/dso/thrust/biosci/brainmi.htm and lays out the following goals:

"1. Extraction of neural and force dynamic codes related to patterns of motor or sensory activity required for executing simple to complex motor or sensory activity (e.g., reaching, grasping, manipulating, running, walking, kicking, digging, hearing, seeing, tactile). Accessing sensory activity directly could result in the ability to monitor or transmit communications by the brain (visual, auditory, or other). This will require the exploitation of new interfaces and algorithms for providing useful nonlinear transformation, pattern extraction techniques, and the ability to test these in appropriate models or systems.

2. Determination of necessary force and sensory feedback (positional, postural, visual, acoustic, or other) from a peripheral device or interface that will provide critical inputs required for closed loop control of a working device (robotic appendage or other peripheral control device or system). Such feedback could be received from peripheral systems or sent directly into appropriate brain regions.

3. New methods, processes, and instrumentation for accessing neural

codes noninvasively at appropriate spatiotemporal resolution to provide closed loop control of a peripheral device. This could include both fundamental interactions of neural cells, tissue, and brain with energy profiles that could provide noninvasive access to codes (magnetics, light, or other).

4. New materials and device design and fabrication methods that embody compliance and elastic principles, and that capture force dynamics that integrate with neural control commands. These include the use of dynamic materials and designs into working prototypes.

5. Demonstrations of plasticity from the neural system and from an integrated working device or system that result in real time control under relevant conditions of force perturbation and cluttered sensory environments from which tasks must be performed (e.g., recognizing and picking up a target and manipulating it).

6. Biomimetic implementation of controllers (with robotics or other devices and systems) that integrate neural sensory or motor control integrated with force dynamic and sensory feedback from a working device or system. The first phase of the program may include dynamic control of simple and complex motor or sensory activity directly using neural codes integrated into a machine, device, or system. Simple actions considered include using a robotic arm or leg to sense a target, reach for it and manipulate it, throw or kick an object at a target, or recognize a sensory input and responding to it (visual, acoustic) directly through input/output brain integration. More complex activity may include issues related to force or sensory perturbation in more complex environments. "

Of 6 goals not once is the identification of the brain's dynamic system mentioned. It is simply and obviously assumed that such a discovery would be a necessary requirement.

And since the scientists at Duke are of the same observational regime as the rest of neuroscience the chance of their stumbling into the dynamic system is all but impossible.

By standard knowledge the discussion of this piece centered in aspects of actual processing relies on that dynamic system and it has undoubtedly caused the aforementioned amusement or disdain by learned readers.

So what are the conditions and requirements of a dynamic system if it were to be used in a brain?

It must be uninterruptible by environment.

Not even super-strong magnetic fields would be permitted to decohere it. Your visit through the fMRI would not render you a vegetable.

It must be flexible by being convertible between charge state in a closed cellular pathway to charge state in a release of chemicals (the same process but instead of traversing a distance able to be subjected to feedback of any interruption in that pathway it has to be able to be converted to a send-receive only condition: like a diode controlling the direction of flow).

The conversion is necessary since the neuron does not process chemicals yet it gets its signals from a charge state induced degree of chemicals.

Therefore it must be able to be present in cellular structure whether singular or in connected tissue.

And most importantly it must be a dynamic system and not a static system.

The signal must be variable enough when used in the brain to cover the complete distance between the near null of its amplitude to past the maximum of the brain's amplitude tolerance.

It must be used for every input and output of the body using the brain regardless of what type of input or output it is as long as it is either biological and therefore carries the charge state or is artificial and duplicates the charge state.

So what is the standard acceptable thought Duke is using to attempt to find this dynamic system? According to Duke:

"The technological revolution of the modern era is changing how we acquire and conceive of biological data. This is due to an explosion in the amount of data that can be acquired with an ever-increasing array of modern tools. The meaningful reduction of data (in the analytic sense) is becoming an increasing priority, and taking on increased urgency. We are developing new signal processing strategies to look at the full spatio-temporal characteristics of human brain data obtained tools like the recording of brain waves (EEG) or the imaging of brain function using functional MRI. An important goal is to translate these signal processing tools developed in the laboratory, to the clinical setting where the problems of data analysis have reached a crisis state."

The author of balony.com best said it:

"...In a nutshell, the universe according to WMAP is 13.7 billion years old, plus or minus 1 percent. It is geometrically "flat," in accordance with the simplest solutions of Einstein's equations, which equate gravity with the bending of space-time. By weight it is 4 percent atoms, 23 percent dark matter - presumably as-yet-undiscovered elementary particles left over from the Big Bang - and 73 percent "dark energy."..." - `bout time to look at tapping that "dark energy" I think"

"Dark energy" is not the same thing as 'dark matter' and 'dark matter' is not necessarily a left over result of any size bang.

Viewed a different way: The universe is a majority 'dark energy' and a severe minority of observable 'non-dark' matter. How then can judgments and deductions of causes be made without making judgments and deductions from what makes up the majority of the universe?

It cannot. It leads to misinterpretations and thereby biases against looking in the right place and looking the right way.

It all depends on how measurement is taken.

In a series of brain related studies numerous scientific examinations have resulted in press releases over the past few years of the standard observational techniques and deduction methodology applied to them:

"People who have trouble controlling their anger might be suffering from a mild form of brain dysfunction, say US researchers. Their tests suggest that being prone to aggressive, even violent, outbursts is linked to impairments in a region of the brain called the orbital/medial prefrontal cortex circuit." [2]

"Psychiatric diagnoses of the condition, known formally as Intermittent Explosive Disorder (IED) are rare. But Mary Best at the Children's Hospital of Philadelphia says: "Even individuals who do not meet full criteria for the disorder can still have frequent uncontrollable episodes of impulsive aggression, and they can impact society through their violent behaviour. An example would be a spousal abuser." [2]

"The disorder usually starts in adolescence, and most cases probably go undiagnosed, says Best. Some psychologists have blamed IED episodes for recent cases of school pupils massacring fellow students" [2]

The condition being observed is one of imbalance between long-term and short-term processing and results in long-term reactionary outbursts, as short-term processing is not being used to control long-term reactions with reasoning.

It is the near extreme condition of simply not using the potential given to the subject. It does start in young ages as the subject is not nurtured into using rational short-term evaluation and in response is controlled by long-term near matches resulting in immediate and prior to awareness actions.

It is the single most important mental cause of criminal behavior.

When the subject is faced with a condition requiring some form of response and is not applying short-term control the result is going to be reactionary long-term motion and since long-term is past history mixed with current stimulus the result can only be predictive behavior and not rational behavior. It also means there is no sense of consequence. So punishment does not deter crime it only comes as part of the sequence of getting caught.

In another study: "A person's brain reaction to a happy face is linked to how outgoing they are, new research shows. The discovery might help researchers learn why some people are more extraverted than others." [3]

"For years, scientists have known that a part of the brain called the amygdala activates in response to fearful faces, and is important in emotional learning and memory." [3]

"But researchers were perplexed to find that, unlike the fearful face response, which consistently activates the amygdala in all people, not everyone's amygdala responds to happy faces." [3]
"Now, research by Turhan Canli at the State University of New York at Stony Brook shows that the more outgoing a person is, the stronger their amygdala's response to happy expressions." [3]

"The biggest impact of this study is the idea that we have to take individual variations into account when studying emotional responses," says Liz Phelps, a psychology professor at New York University." [3]

Another difference between the use of long-term reactionary processing and the degree to which short term rational processing controls it:

"Whether amygdala activity helps cause extraversion or extraversion causes amygdala activity is unclear. "It's impossible to answer that with the data we have right now," Canli says." [3]

"Still, he speculates that researchers may someday explain extraversion and other complex personality traits based on the responses of brain structures like the amygdala." [3]

Amygdala processing is where long-term and input come together and result in new seeded memory wavelets for short-term processing which results in seeding long-term memory (in humans).

When short-term processing is not being used to at least the degree of its potential that equals the output of long-term processing, long-term processing is going to result in more output than it should in a human.

Those dominated by long-term reactionary memory processing are going to be introverted. Those exercising a degree of short-term memory processing in excess of the balance with long-term are going to be extraverts.

Which is why most criminals are introverted and dominated with prejudices, misconstrued blame and retributive tendencies.

In another study at the University of Pennsylvania researchers deduced that "Brain scans can reveal whether someone is lying or telling the truth, US researchers have discovered. When people lied, fMRI (functional magnetic resonance imaging) scans revealed significant increases in activity in several brain regions." [4]

"Daniel Langleben and his colleagues at the University of Pennsylvania hope fMRI could be used for more accurate forensic lie detection. The widely used polygraph test is based on changes in heart rate, blood pressure, breathing and the electrical resistance of the skin. But these factors can vary widely among individuals, making it more difficult to establish whether someone really is telling the truth. " [4]

The polygraph test measures long-term inhibited responses and attempts to deduce that if long-term inhibitory response is being curtailed (short-term memory is active to evaluate and reason the parts of the input) then it must mean a person is lying.

How absurd. It means a person is thinking in long-term

which could mean a person is rationalizing in short term which is effecting long-term and not making sense to long-term so outputs to motion including those things measured by the polygraph are receiving signals that are not normal and therefore result in increase activity of the measured outputs.

"Langleben's team gave 18 people an object to hide in their pockets. They were then shown a series of pictures, including one of the object itself. As each picture was presented, the participants were instructed to deny that it matched their hidden object." [4]

Input is presented to long-term memory where outputs are given to motion including those things monitored by a polygraph. Short-term is where rationalization happens which means a person being told to lie is both contemplating the act as a lie and seeking to rationalize the presentment of a lie.

"When there was a match, and the person was lying, activity in several regions increased. This included the anterior cinglate, which is associated with response inhibition and error monitoring, and the adjacent right superior frontal gyrus, which plays a key role in attention." [4]

There is no area of the brain that specifically handles error monitoring. The entire computation process can result in error based on input or memory content. When input does not relate to long-term memory the result is uneasiness manifested in "changes in heart rate, blood pressure, breathing and the electrical resistance of the skin".

"The results suggest there is a "localised brain correlate of deception", the team says." [4]

The result suggests there is a long-term process not matching a short-term evaluation and it may have nothing to do with the topic being studied.

If the subject is using a degree of short-term processing in advance of the balance with long-term output then that person is going to evaluate far more criteria than the simple test permits study of.

The result is highly intelligent short-term aural thinkers and highly intelligent short term visual thinkers will be construed to be lying when in fact they are thinking because input causes relationships to be compared in long-term, outputs to short-term where the comparison is reasoned. The more reasoning going on the more potential little connections will not make sense and cause more deductive reasoning and more accusations of lying in such a test.

The polygraph is a nice long-term monitor and absolutely no indication of truth.

In a study called "**Mind theory**" [5] Francesca Happe of the Institute of Psychiatry in London deduced: "People with autism lack this "Theory of Mind", and show abnormalities in these three key brain areas." [5]

"Pinpointing brain dysfunctions involved in autism should lead to a better understanding of the causes of the disease. It should also help to develop accurate methods of evaluating autism treatments and to investigate whether animals have a Theory of Mind, she says." [5]

By "theory of mind" she means a sense of self. Autism is a disease of the feedback loop or return pathway of brain memory in short-term. The same condition exists in long-term but is referred to as memory loss or amnesia.

In short-term loss of the return ceases the loop of awareness and causes the individual to be completely at the mercy of their dominate type of thinking.

Visual autism results in a near stupor where motion is repetitive and loops at the level of existing remaining return while aural autism results in being unable to a degree to connect attempts at gaining the subjects attention with being the subject of attention.

Without a looped (and deep enough to make a difference) short-term memory inputs are passed from long-term output to short-term processing to long-term memory without much rationalization and without the ability to set up a loop in long-term of being aware of being aware.

"These individuals use different brain areas to solve the problems - it seems that they are using sheer intelligence rather than an innate social intelligence," Happe told **New Scientist**." [5]

It would appear that not only is intelligence not defined but the use of the terms 'sheer' and 'social' are observational illusions.

Visual dominance would result in an observation of social while aural dominate would result in the observation of 'sheer'.

"But activity in these three areas did not increase in the Asperger's patients while they completed the same tasks. "Instead they activate other areas of the frontal lobes. This fits in with the idea that they solve social problems through general intelligence," she says." [5]

Asperger's is the name given to the low interference form of the same malady responsible for autism. Instead of severing the short-term loop completely as in full autism Asperger's subjects suffer from selective interruption making the sense of self less intrusive in long-term looped memory.

In one of the more comical research studies (a prime candidate for the poster child of observational illusion) researchers at the University of California, Santa Barbara deduced "Part of the human brain is dedicated to detecting cheats." [6]

Yes, of course. Cheating is something brains can do so there must be a cheat place in the brain.

(That was sarcasm.)

"We think it develops in all normal individuals, and that it develops in part because our brains were selected to develop this competence," says John Tooby." [6]

"Tooby and his colleagues studied a man who suffered accidental damage to the limbic system, a brain region involved in processing emotional and social information. RM, as he is referred to, performed as well as other people on one set of reasoning problems, did much worse on problems specifically designed to test reasoning about social exchanges." [6]

"At its simplest, social exchange runs along the lines of "you scratch my back and I'll scratch yours". Previous work has shown that people, and some animals, are extremely good at keeping a check of who owes who within a group - and at spotting and punishing cheaters." [6]

"Researchers had proposed that general reasoning abilities could account for this. But RM's deficit suggests that detecting social cheaters depends on specialised neural circuitry, the team says." [6]

"Their conclusion is "robust," says Nigel Nicholson, an evolutionary psychologist and director of the Centre for Organisational Research the London Business School. "It's essential we have trusting relationships with people in communities where we are highly interdependent for survival and reproduction. Cheat detection is very important," he adds." [6]

"RM recorded a score of 70 per cent on the precaution rule tests - the same as the controls. But he scored only 39 per cent on the social contract tests, compared with 70 per cent for the non-brain damaged people." [6]

"Identical tests on two other people with brain damage similar to BM's, but with a slightly different pattern of damage, showed that their social contract reasoning was unimpaired." [6]

"RM's differential impairment indicates that being able to detect potential cheaters may be a separable component of the human mind," the researchers conclude in the journal *Proceedings of the National Academy of Sciences*." [6]

Here is the rub: "However, if a region of the brain has evolved to specialize in cheat detection, it should be present in all people, the team reasoned. Most experiments are performed on people living in modern, western societies." [6]

So they looked at other societies.

"What is quite amazing about their performance on cheater detection is that it flies in the face of all ordinary ideas about learning a higher level cognitive skill," Tooby told **New Scientist**. "People are just as good at utterly unfamiliar rules as they are with rules that are personally and culturally highly familiar." [6]

It is long-term memory that presents the history of a subject's short-term rational deductions. It is the short-term processing of memory and the return of it in a complete closed loop system that provides 'social contact reasoning'.

The researchers observed two forms of processing and deduced that the results of a brain functioning as it does somehow starts with the brain function as it results.
How absurd. But that is the way AI research works. Observe something agreed to be the result of an 'intelligence' then find a model that will result in the same thing and call it intelligent. It is not. It is just another way to reach a similar conclusion and without using the mechanism of intelligence is nothing but a nice physical interpretational painting in code.

On the surface a study regarding chewing gum and brain function may appear to be a bit elementary minded. In reality it shows a function of the brain not even considered by the researcher's deductions:

"Chewing gum can improve memory, say UK psychologists. They found that people who chewed throughout tests of both long-term and short-term memory produced significantly better scores than people who did not. But gum-chewing did not boost memory-linked reaction times, used as a measure of attention." [7]

Tests of long and short term are completely unable to separate the two unless damage has been found in the brain severing long term memory creation or long term memory output to short-term.

"These results provide the first evidence that chewing gum can improve long-term and working memory," says Andrew Scholey of the University of Northumbria in Newcastle, UK. "There are a number of potential explanations - but they are all very speculative." [7]

One third of the 75 adults tested chewed gum during the 20-minute battery of memory and attention tests. One third mimicked chewing movements, and the remainder did not chew. [7]

The gum-chewers' scores were 24 per cent higher than the controls' on tests of immediate word recall, and 36 per cent higher on tests of delayed word recall. They were also more accurate on tests of spatial working memory. [7]

"The findings are intriguing, although it is clear that questions remain to be addressed," says Kim Graham of the Medical Research Council's Cognition and Brain Sciences Unit in Cambridge, UK. "In particular: what is the mechanism by which chewing improves memory?" [7]

"There are three main potential explanations, says Scholey. In March 2000, Japanese researchers showed that brain activity in the hippocampus, an area important for memory, increases while people chew - but it is not clear why." [7]

"Recent research has also found that insulin receptors in the hippocampus may be involved in memory. "Insulin mops up glucose in the bloodstream and chewing causes the release of insulin, because the body is expecting food. If insulin receptors in the brain are involved in memory, we may have an insulin-mediated mechanism explaining our findings - but that is very, very speculative," Scholey says." [7]

"But there could be a simpler answer. "One interesting thing we saw in our study was that chewing increased heart rate. Anything that improves delivery of things like oxygen in the brain, such as an increased heart rate, is a potential cognitive enhancer to some degree," he says." [7]

"But a thorough explanation for the findings will have to account for why some aspects of memory improved but others did not, Graham says. She points out that gum-chewers' ability to quickly decide whether complex images matched images they had previously been shown was no better than the controls'." [7]

The answer is the ratio-enhanced condition of brain processing.

Chewing gum for most avid chewers takes place at a normal or near normal 1 to 2 times per second. Try it. 1 per second and you are casually chewing gum using long-term memory output. 2 per second and you are showing a contemplation going on in short-term memory.

Long-term memory output is the same 10 per second (average) as short-term as both outputs merge to the same actuator.

But the processing speed of long-term memory is closer to 30 times per half second or around 60hz. It can be observed that supporting with feedback of motion set to replicate at the processing speed is going to serve to stabilize the processing speed and synch it with the input speed making each sample a memory wavelet of a near equal distribution of the input event.

That is mostly felt in aural processing but only because it is easier to display with aural processing.

Aural (both long and short term) outputs to the head: Mouth, tongue, jaw, facial expressions, speaking: and since it has recorded patterns of such repetitive balance before while listening to music recognizing a pattern is much easier to accomplish and the result without short-term control is a near zombie like replication of that pattern. It is what forces people to their feet on dance floors.

Chewing gum supports that pattern of 'beat' and since it so closely matches the actual processing ratio-enhanced speed of long-term the results are going to be based in long-term processing.

While studying the result of the operating system of the brain, which is working within biological hardware working the same way each time it is applied the concentration of research deductions, has been that hardware.

As is evidenced by the DARPA study with Duke there is a strong belief that there is an actual dynamic system operating within the brain's hardware.

The logical deduction from that concept is that hardware is dedicated function mechanics but it would appear researchers are in the mode of seeking mechanical solutions to unexplained observations and wind up ignoring the system being processed in the

40

mechanics.

The 'software' or function system using the hardware is the result of the hardware's ability to do the work without corrupting it.

Improper wiring will corrupt as well as cellular deficiencies, reduced or increase base amplitude and even repetitive incorrect inputs.

A fine example of how mechanics can get in the way of good science is indicated by the study of dreams presented by Robert Stickgold of Harvard Medical School. [8]

"Asking people with amnesia to play Tetris has revealed clues to why dreams are so illogical - and perhaps to why we dream at all." [8]

"Most sleep scientists think that in our first dreams of the night we access "declarative" memories of recent events - those we can consciously recall. People with amnesia cannot form this kind of memory because of damage to the hippocampus, a part of the brain." [8]

"But Robert Stickgold of Harvard Medical School found that people with amnesia who had no recollection of having played Tetris still "saw" the computer game's falling blocks in their first hour of sleep." [8]

"He thinks this shows that dreams during so-called Stage One sleep rely on abstract, subconscious memories, which amnesiacs are still able to access. And he thinks this explains why dreams often seem disordered and bizarre. " [8]

"When I saw the results I almost fell of my chair," Stickgold says. "We thought that if there's one part of sleep that depends on declarative memories, which amnesiacs lack, it's sleep onset." [8]

Dreams seem disoriented and bizarre due to the dynamic system's connection processing.

Long-term is always being compared to input.

When input is nearly stable and therefore either looped itself (as in day dreaming) or nearly shot down (as in sleep) memories are being compared to either the same thing or to nearly nothing which is also the same thing.

That allows returning memories to pass through without relative time relationships.

Seeing a red ball before going to sleep results in the 'red ball' event in all of its various orchestrated perspective degrees of memory in separate processing pathways mixing with other memories that are not in the same time frame which makes 'red' connect to anything else 'red' in memory while 'ball' is connecting to everything else 'ball'.

Red is a color, a specific color, potentially the color of a must feared image or even the misinterpretation of the audible sound of the word and relates to having visually consumed words (read).

Ball is a shape, an object. It will relate to all things round, rounded, singular in shape, soft angles and when mixed with the most fear image in the subject's long-term memory will cause the visual dream of a horrible variation of a known fear.

Stickgold is not at fault for this misinterpretation of brain function he is only following the same misinterpretation brought on by observational illusion.

"Declarative memories "are ones you can declare you know: 'I had eggs for breakfast' or 'My brother's name is Ed'," Stickgold says. "Implicit memories are ones that are in your brain but you can't access consciously." [8]

This statement is so misaligned it is nearly laughable.

Memories you can declare to know are actually memories either still within short-term or memories that are not repetitive. I had what for breakfast? Remembering a standard event is not an easy thing. But it is still in upper level depths of long-term memory even after it is no longer focused by short-term.

'My brother's name is Walter.': (well, I'm saying it now), is not at all a recent memory. That memory goes back as far as the knowledge of Walter existed. Its residence in upper level long-term will only cause the previous memory in long-term to be supported and it will remain higher amplitude longer.

For the new memory to be its own potential long-term loop (recognition) it would have to be in an environment that creates it own loop and not simply a reference by auditory processing.

"Stickgold thinks the access to implicit memories that dreams give may be essential for learning. "One of the most difficult problems the brain faces is how to put together information from different sources, to see how things fit together." [8]

It is good to learn (and you are if you have read this far) but the brain does no such thing.
Not only is it not difficult it is mandatory.

What I believe he meant to say was that learning what causes dreams putting together different memories and how they fit together to create dreams is important to learn.

That I will agree with. But it just gets deeper:

"By blocking declarative memories and forcing the system to work with these weak associations, the brain is coerced into looking for unexpected, novel and potentially highly creative and useful connections that otherwise we would not notice." [8]

The brain cannot block short-term processing short-term processing can block its own use but one can only affect the other the way it is wired together.

The brain cannot be coerced. It can be tricked such as saying 'we would notice' is a declaration of such 'noticing' being real. It is not.

It is a result of previous 'notices' mixed with the current notice and controlled by the type of thinking employed by the subject and the degree to which they exercise the potential short-term processing ability they are born with and not corrupt such use if indeed it is being used with long-term memories based in improper assumptions.

""The study has shown that the mental activity at sleep onset is clearly related to the learning situation," agrees Carlyle Smith, a sleep scientist at Trent University in Peterborough, Ontario." [8]

That indeed is true.

So is every moment of sleep a replication of being awake without the benefit of the stability of placing input as a controlling factor to know what the environmentally imposed perception of reality is for the subject.

Including other more assumptive assertions the study's press release or perhaps the New Scientist's writer surmised:

"This is particularly interesting," says Richard Haier of the University of California in Irvine. "It suggests that people who learn the best may use the early dreaming the least for learning." [8]

People who learn are living people without damage to the brain in parts of the architecture that effect their observed malady. People who learn the best are a product of either advanced long-term processing, advanced employment of short-term processing or both.

"The resting membrane potential of a neuron is about -70 mV (mV=millivolt) - this means that the inside of the neuron is 70 mV less than the outside." [9]

Which indicates: that while at rest a neuron's state is not charged.

"The action potential is an explosion of electrical activity that is created by a depolarizing current. This means that some event (a stimulus) causes the resting potential to move toward 0 mV. When the depolarization reaches about -55 mV a neuron will fire an action potential. This is the threshold." [9]

The mixture of two opposing charges (the first charged condition amplitude) at the presence of a second charged condition amplitude cause the neuron to perform a mean sum calculation of the two amplitudes which is the variable value transmitted as a result of that 'firing'.

The process that increases the deficiency of charge state to reduce its value shows the presence of a computation.

That computation is a variable-amplitude of the wavelet generated. That wavelet will be transmitted through the axon and start the process over again.

It will remain stable as long as the base amplitude used to start the process either remains at all (the subject is alive), is reduced due to stress reduction or illness not present at all (the subject is deceased.)

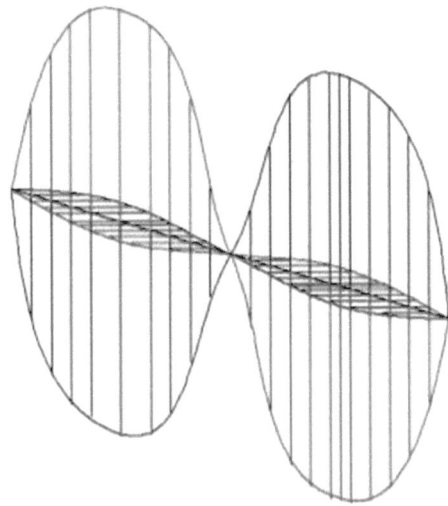

On the left of this figure is the positive charge state indicated by red which contains a singular anti-positive charge while on the right a blue indicated anti-positive charge state holds a singular positive charge.

If the two do not exist together the subject is deceased.

It is so evident in Heisenberg's Uncertainty Principal that it is impossible to measure one state and be able to measure its other state.

Uncertainty dealt more with wave versus particle observation but it has a great deal of application in all other areas of measurement as well.

Working in reverse of positive charge the anti-positive 'dark energy' charge is highly viable and flexible within the confines of its opposing positive energy charge condition.

The equation to simulate the action of a neuron is the same equation used to denote interaction of amplitudes in all other systems:

$$a_z = \frac{a_x - a_y}{2} + a_y$$

This equation translates to a simple average (the average is not used because the algorithm should match the connectivity):

$$a_z = \frac{a_x + a_y}{2}$$

Description of the algorithm to match the connectivity requires that the algorithm display the function as well as the order of function in one depiction.

$$a_z = \frac{a_x - a_y}{2} + a_y$$

Is described:

·   Neuron amplitude output ('discharge') is equal to the joint presence of two amplitudes (initial charge plus 'firing' clock pulse amplitude) acting upon each other's amplitude value (remember, this is discussing the inverted or anti-positive charge state) upon the presence of the second amplitude value.

This computation takes place exactly the same way in all neurons throughout the subject's nervous system.

For such an anti-positive charge condition to be known to exist (empirical versus conceptual) it must be able to be measured.

In the standard connectivity of an electronic measuring device the ground or 'negative' measurement input must be connected to a ground or negative source.

That leaves the positive measurement input to be connected to the location of desired measurement.

Uncertainty: "The more precisely the POSITION is determined, the less precisely the MOMENTUM is known." Is applicable to measurement techniques.

By measuring for a specific thing (property) one has determined the precise position of measurement. Therefore the other thing (property) is unable to be measured. It has everything to do with the method used to measure and nothing to do with the thing being measured.

Of course that is not the commonly held belief system surrounding Heisenberg's statement used as a foundation for the Copenhagen Interpretation. But it is the right description nonetheless.

To measure the Neutronic charge condition the connectivity of the measuring device places the positive measurement input at the positive power location and the negative measurement input on the location of the desired measurement.

It is simply upside down.

The diagram to show this in action:

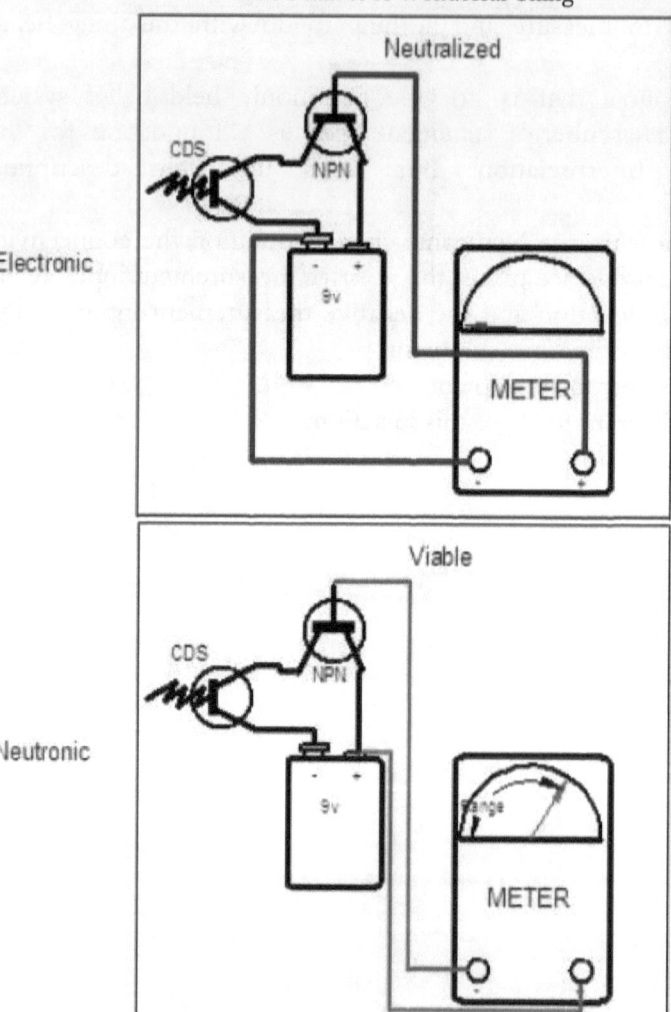

Connecting the VOM as indicated by the Electronic model will show the electronic volts present which will indicate in the −mv up to 0 range.

Connecting the VOM as indicated by the Neutronics model will show electronic volts present that will indicate in the range of the power source's stored voltage down to 0. Combine the two and the result is an electronic voltage that moves from the −mv range to 0 while the Neutronics voltage will read in reverse.

Only the Neutronic voltage is variable and stable and

environmentally secure but it requires the presence of both charge states for either to exist.

In the basic model of the human brain… feedback return pathways and memory computations taking place in neurons using the above regime result in 'firing' as the base amplitude of the host body is applied to each calculation in memory causing the 'threshold' to be reached and a 'firing' to take place.

This process results in two things:

The Neutronic frequency is not affected as amplitudes of the same frequency are joined. This causes no friction and no heat but the present Electronic frequency minimum charge state does experience friction and heat and causes the momentary existence of a resulting voltage past the 'threshold'. Observed as voltage 'spikes' the deduction was that the event was a singular firing and the common belief set up that the condition range for a neuron was either on or off.

While that observational illusion was being proven over and over again yet still did not answer the question of a true dynamic system at work the un-measured Neutronics amplitude convergence process was taking place in a highly variable state.

Neurons are 'seeded' or charged with a memory wavelet amplitude and 'fired' when the presence of the base amplitude is injected by the pulsed clock pathway controlled by the divisions to the biological clock's host wavelength.

In this graphic we start to understand the concept of frequency division or ratio-enhancement (RE). We are viewing the Neutronics wavelength. The Electronic wavelength is present as well but only in the 'on' condition (minimum charge necessary to remain a charge):

One wavelength (established by the base provided by the host's combined Electronic-Neutronic charge state) is presented to the first level of RE representing input firing rates. Each wavelength is thereby reduced from one wavelength to two causing firing rates for input receptors to be working twice per second. (This varies specifically by individual genetic makeup of the biological clock).

Input Receptor Clock Rate

Each half-second pulse is computed by the second reduction or frequency (RE) by a factor of 30:1 to amplitudes running at that long-term memory making amplitude pulses equal in frequency to the reduction

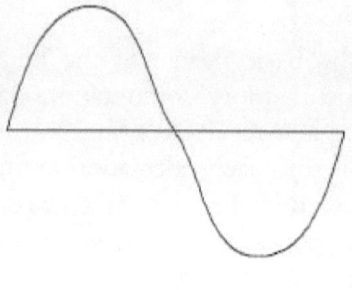

then level of division

RE in output

rate.

(This varies as well in individual subjects based genetic makeup of the clock and accounts for the

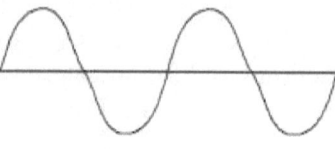

on the biological degree of

'innate' or 'instinctive' behavior displayable by the subject (feedback return pathways also play a major role in the degree of recall potential)) and is the data sent in discharge to both short-term processing and direct long-term memory output.

Each 30:1 memory wavelet from long-term and input comparison calculation is then split again by a factor of 30:1 (Varies widely and accounts for the degree of potential intelligence able to be presented by the subject.

(Feedback return pathways also play a role in the degree of awareness potential))) the data sent in discharge to both long-memory and direct term memory output.

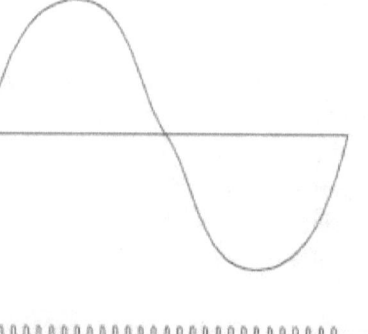

major

and is

term short-

Short Term Memory Rate

Clock

After processing term a reduction takes without specific additional wiring A sample rate firing at second coming from the

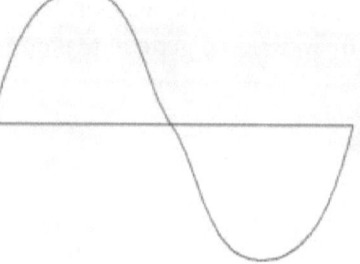

in short-place

required.
10 per

biological clock causes each 30:1 frequency pulse division to be combined into 1:5 wavelengths.

The 10 per second pulse is the output speed of motion actuators (varies by genetic makeup of the subject's biological clock.)

Output to Motion Clock Rate

The ratio differences are caused by division rates of the base frequency (which is an inverted Neutronic frequency with complimentary Electronic frequency present which is the wavelength used in wave type definition.):

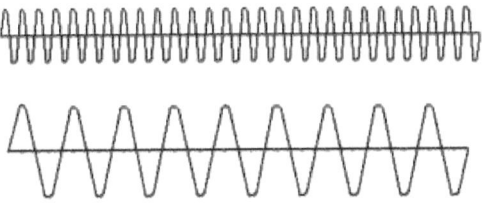

Base Frequency: 35,721hz causes pulse at 1hz Delta (0.1 to 3 Hz)

Input Frequency: 35,721hz / 2 = 17,860.50 hz causes pulse at 2hz but is blended by the Long-Term processing rate in distributed clusters so as to generate a very slightly higher harmonic. Theta (4-8 Hz)

Output Frequency: 35,721hz / 10 = 3,572.10 hz causes pulse at 10hz Alpha (8-12 Hz)

Long-Term Frequency: 35,721hz / 60 = 595.35 hz causes pulse (30 average oer half second) at a wide genetically controlled range of frequencies within the Gamma range.

Short-Term Frequency: 35,721hz / 1800 = 19.845hz causes pulse (900 per half second) at a wide genetically controlled range of frequencies within the far upper reaches of Gamma.

Clustered distributed near-rate firing sequences can cause harmonic results and are listed in the Beta (low, mid and high range) wave type.

Clustered distributed firing rate sequences can cause harmonic results from the abundance of both long-term and short-term memory neuron pathways so as to result in a singular closer range observed in Gamma presence.

Keep in mind that there is just as much process of the clock's pulse pathways as there are memory and computation pathways and they blend together as well to form harmonics that are generally observed to be the four types of brain waves.

How the architecture accomplishes this division of frequency to cause increase of frequency processing is one of using the same regime of connectivity as the return feedback pathway.

The following graphic shows the method of dropping frequencies by hardwiring:

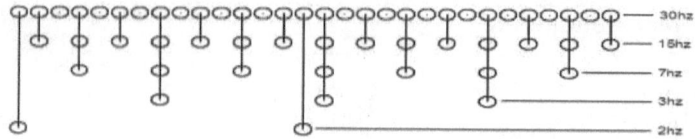

In memory feedback return pathways the same divisional process is used to return memories in the same order of time:

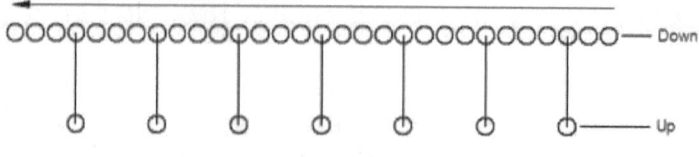

The ratio of pathway U-Turn is directly proportionate to the recall clarity ability of the subject.

Most importantly there is literally no other way to mechanically wire a process that requires recent to be compared to recent and only compared to not so recent if the subject retains the input enough to receive deeper similar connections.

Applying both the dynamic system of processing (the architecture that makes it possible) with the application of pulsed differences in processing speed by level of the system and the result for the human brain is this:

The Human Brain

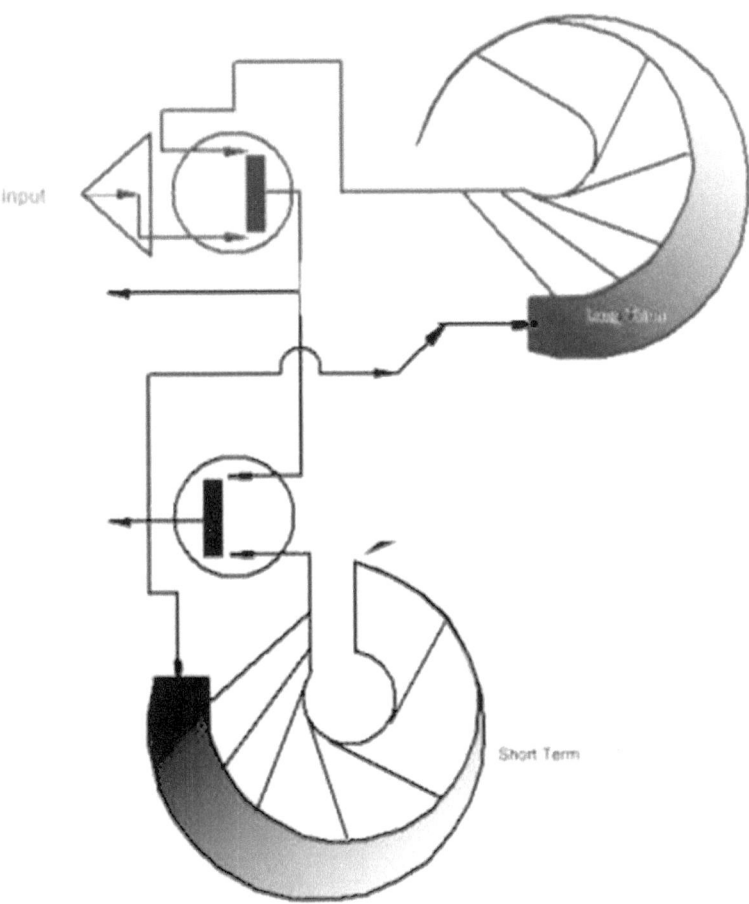

The brain is a wonderful thing.

It gives us the ability to see, hear, sense pressure and temperature as well as taste flavors and appreciate smells, remember, recall, evaluate and deceive ourselves that *we* are doing those things.

Chapter Three
The Top Questions About The Brain

On June 10, 2003 the call was made for questions in numerous online forums. 10 days later the call ended:

Moving from **step one** to **step two** implies completion of step one.

For most of our collective human history **step one** of understanding the brain has been routed in ignorance.

Ignorance is a horrible thing.

It is not surprising that the first attempts at explaining the brain began where they did:

Nicolaus Steno, born in Copenhagen in 1638 as "Nils Stensen, latinized to Nicolaus Stenonis, and anglicized to Steno" [4] is quoted from 1669 as saying:

"The brain, the masterpiece of creation, is almost unknown to us." [1]

Until the time that science had managed to actually delve into the physical brain a great deal of speculation ruled the study:

To Aristotle the brain was not the primary organ of the body. The heart was. He coined a term:

"*sensus communis*".

We use it to this day (common sense) although not in the way he meant it. (Common-Sense is defined and illustrated in chapter twelve.)

"By the first century A. D., Alexandrian anatomists such as Rufus of Ephesus (50 AD) (in "On the naming of the parts of the body") ha d provided a general physical description of the brain.

Basic structures such as the *pia mater* and *dura mater* (the soft and hard layers encasing the brain) were identified in addition to the basic divisions of the brain itself.

Building upon this research in the next century, the Roman physician Galen concluded that mental actively occurred in the brain rather than the heart, as Aristotle had suggested. His observations of the effects of brain injuries on mental activity formed an important practical basis for his conclusions.

Galen concluded that the brain was the seat of the animal soul – one of three "souls" found in the body, each associated with a principal organ. The brain was a cold, moist organ formed of sperm." [1]

"In the Middle Ages, the anatomy of the brain had consolidated around three principle divisions, or "cells," which were eventually called ventricles. Each cell localized the site of different mental activity.

Traditionally imagination was located in the anterior ventricle, memory in the posterior ventricle, and reason located in between. Yet where was the *sensus communis*?

The Islamic medical philosopher Avicenna wrote in the early eleventh century that it was housed in the "faculty of fantasy," receiving "all the forms which are imprinted on the five senses." Memory preserved what common sense received.

By contrast, the great anatomist Mondino de' Liuzzi wrote in his *Anatomy* (1316) that common sense lay in the middle of the brain. Aware of the contractions that had proceeded him, he affirmed that "there is only the *sensus communis* which is variously called fantasy and imagination." "[1]

"Other problems remained open to debate.

For instance, Avicenna chastised physicians for favoring Galen over Aristotle. A century later, Master Nicolaus of Salerno marveled at the confused humoral accounts of the brain. "The brain ... is, according to some, of hot complexion; according to others, cold; according to others, moist." Such difference of opinion underscore how little was known of the brain's anatomy, let alone its physiology." [1]

It also underscores a pattern of brain study.

Whatever was in 'vogue' at the time was the reference used.

Today we reference the brain in terms similar to the only other thing we know of that looks like it does the same thing. The computer.

A computer is logical to us because we understand the logic it uses. A computer is a thing that takes space and as such a thing it has a location we are aware of.

A computer has sequence in that its functions are understandable in order. A computer has size and we all know bigger or faster is better, right?

As humans we all strive to find order. Sense. Logic. Placement. Sequence. Size. Importance. observations of the effects of brain injuries on mental activity formed an important practical basis for his conclusions.

We use the measurements in vogue to measure ourselves and we impart a kind of first step stumble along the way.

We keep coming back to the first concept made.

That concept was location.

It is our sense of time that gives rise to the need for order and our need for order that gives rise to the notion that time somehow exists outside of what gives its awareness to us.

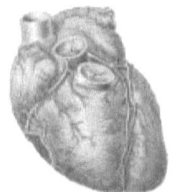

At first, the brain was considered another organ but not **'the'** organ since the **'heart'** was 'the' organ of the day.

Once we understood the heart was a nice pump science stopped calling on it to think but religion held on to the idea that the heart was something more.

Over the years the single most prevalent concept employed in brain study has been location.

We have moved from long held beliefs that thinking was in the heart to the now long held belief that the brain has 'centers' of subjective observational causes.

We have moved from 'centers' of the brain to 'genes' of the DNA.

Almost every month we witness new claims from under-nourished researchers having found a gene that is responsible for some subjective observable trait. We never hear the follow up. And when asked about the follow up the press shrugs it off.

There is a 'doctor' now hitting the talk circuits, speaking to sports teams and being reported in newspapers claiming there are 16

different types of brains. When the reporter was informed of how ridiculous that claim is the response was this:

" This wasn't meant to be an investigative piece, more of a fun/feature.

There is no doubt many in the scientific community disagree with this gentleman, but there is also no argument that major sports teams are paying him big bucks regardless."

That response was received from WFIE at http://www.14wfie.com

It doesn't matter that the whole concept is seeing symptoms and developing uses for them, at least not to the news.

As our progress in brain studies has moved ahead concentrating on **'where'**: we learned that **'where'** has ... parts.

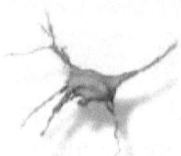

And we all know that reverse engineering means looking for something smaller as a cause. Or do we?

After all of the studies, surgeries, autopsies, dissections and now DNA research we are still consumed with where and what part is responsible for what thing we can observe.

We can observe many things the brain does (since now we know the brain is doing them) but without language we could not communicate those things to others.

Language applied to new things needs comparisons by which the new thing can be judged to make it easier for students to comprehend.

Fancy that. Comparisons are used as a matter of course to make things more understandable.

Just like the brain itself, when a comparison is used enough it is no longer comparable to the original goal it is comparable to the comparison it is.

Language has provided us with the ability to impart our individual perspective upon others who, given enough reason to trust what we say (whether that is by our title, our affiliation, our publisher, our name brand) will either accept our perspective as theirs or impart their own perspective on ours to create a hybrid and oft' times convoluted observational deduction.

When others have attempted to answer questions about the brain they do so by giving its parts and describe what they are. Check [2] for one of the most interesting such examples.

"There are certain phenomena which once seemed familiar and obvious and appeared to provide an explanation for things which had been obscure.

Subsequently, however, these phenomena began to seem

quite mysterious themselves and began to arouse astonishment and curiosity.

These phenomena, above all others, were zealously investigated by the great thinkers of antiquity." [49]

Physicists on the other hand address the brain from a more 'human' approach. 'Human' approach means by observation.

The empirical. Physicists call that qualitative properties, which involves quality of any kind and is essentially and inherent feature or Qualia for short.

Qualia, when evaluated, is essentially what Merriam-Webster says it is: "a property as it is experienced as distinct from any source it might have in a physical object".

That would make the study of Qualia subjective. "Characteristic of or belonging to reality as perceived rather than as independent of mind". (MW)

Physics is not alone in attempting to explain the 'black box' by looking for things 'black'. Psychiatry and Psychology do the same thing.

If apples come from apple trees how can the duplication of an apple artificially result in understanding the tree? It cannot. There are many ways to duplicate anything. There is only one way to make it for real.

If we referred to our cars in the same way we refer to intelligence we would all be driving the latest artificial horse because transportation 'evolved' from horse to automobile yet we don't.

We do still refer to its power as a horse.

Horses replaced human legs for more efficient transportation yet we do not call horses, artificial leg devices.

How we 'view' things (not from which variety of the 16 different brain types we are supposed to use, according to the over paid motivational speaker with an angle) depends on how our brains are wired.

 L. Ron Hubbard was a highly visual short-term thinker who like most other persons never gave it a thought that others may not think the same way.

He 'discovered' "...that the mind was, in fact, a collection of mental image pictures." [3]

For some people, yes, for others, no and still for others yes, and no.

L. Ron Hubbard was a highly visual short-term thinker who like most other persons never gave it a thought that others may not think the same way.

Hubbard then claimed that the 'mind' "is capable of taking 25 pictures per second. At that rate, the mind is recording 1500 pictures a minute." [3]

And "...*Dianetics* explains that it's not just the visual record — but many more perceptions are recorded, as well: temperature, smells, sound, mood, heartbeat, muscular tension, motion, and dozens of different perceptions.

As Ron later discovered, there are 57 different perceptions.

Multiply that by the number of mental image pictures you record in sixty seconds, and that's nearly a hundred thousand concepts committed to memory in one minute!" [3]

But wait, we're not finished yet:

"Each mental image picture is made up of millions of bits of information. So, the mi nd is taking in trillions of bits of information every hour." [3]

The assertion that 'pictures' are in the 'mind' is indicative of the time it was thought.

1950 was the height of motion pictures as a phenomenon.

It was the beginning of television and the start of the growing visual culture but the brain no more takes pictures than it records tapes of sound.

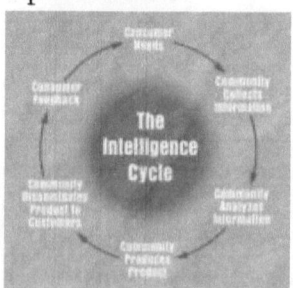

The medium we see is not the medium that sees.

The perceptions recorded by the brain are not the perception we have of it.

Hubbard was right in his assertion that memory is recorded in order but we do not need a brand name "time-track" to explain memory and using time to explain memory is like eating an apple and explaining the process that allowed the tree to give fruit.

By observing what comes out of the brain our science has tried to duplicate those things observed as 'intelligent'. But what is intelligence?

The term 'intelligence' is applied to

knowledge gathering by 'intelligence services'. But is intelligence the gathering of knowledge? Do you get more 'intelligence' when you gain knowledge?

The United States Government Printing Office provides an intelligence cycle of handling information (knowledge). But is handling information intelligence?

The Cray Super Computer is a pretty powerful piece of hardware. It handles information pretty darn fast. But it isn't the slightest bit intelligent.

So what is 'intelligent'?

"**Intelligence** is a general mental capability that involves the ability to reason, plan, solve problems, think abstractly, comprehend ideas/language, and learn." [50]

That definition would make intelligence the cause of what makes a brain intelligent.

Nice definition for not having to be very specific.

And it fits the comfortable 'mind-set' of those who continue to duplicate, treat, diagnose and 'mess-with' what they have not a clue about. Anything can be justified if one does not have to define it.

Intelligence is actually the process that gives rise to observable intelligent behavior. It isn't a very specific definition but it is not its own conundrum either.

By defining intelligence as the process that gives rise to observable intelligent behavior we are able to address the process and the system from which it stems and are able to delve deeper into that system for what causes all other observable brain results as well as some not so observable yet readily available.

To have the slightest chance at being accurate any such examination of a process must not be cluttered with the subjective. After all, the method of science is supposed to be objective. A process can only be objective or it is not a process it is a result of a process.

BRAIN 101:

Luckily for us the two main processing pathways of the brain are both located in the same close proximity area.

The head.

Can you imagine if the two competing processing pathways were pressure and temperature?

We would perceive 'I' to be 'all' instead of 'I'.

Our mind needs order and placement just as much as every other process of intelligence.

That we do not perceive other things the brain is known to cause or be involved in causes us to label locations of activity we associate with those subjective experiences and we are satisfied without objective reason.

We, as humans perceive ourselves to be in our heads.

There are those who do not perceive a placement of the 'mind' at all.

Those people are subject to want to achieve such a 'thing' if its form of presentment provides incentive to fulfill some other need or desire or subject to reject such a thing if it challenges an already accepted truth.

The mind is the name we have given to the 'thing' that is us, the being that we are. It has been called other names in the past and sought after through every conceivable attribute given to it from subjective evaluation:

· The soul: we know we better protect it as it is us and the thought of anyone or anything else 'possessing' 'it', even if we do not perceive ourselves to actually 'have' one is a horrible contemplation and easily capitalized through secret possession of mystical means to ward off such 'possession' and if we do not perceive a 'mind' yet desire to find fulfillment we may be tempted to fall for that.

· The enlightenment: the level of being that we can become by following someone's clear cut path that is able to be obtained personally without another person's method or rules and therefore implies lower levels of being and since only the few and perhaps chosen have attained such higher level of being those not yet doing so are relegated to lower levels of life as well as enlightenment.

· Meditation: the word given to the process of reaching inside to become one with the power within. If we do not perceive such 'power within' yet seek identification of the 'thing' we know is in there we might be tempted to believe that acquiring the 'power' is fleeting and only for the very wise.

• The 'power': the name of the universal spirit thought to justify our existence as we could never have the 'power' unless we reach a level of 'understanding' and is acquired through either prayer or sacrifice which is in some beliefs described as being veiled by the 'illusion'.

• The spirit: where a connection is made to the potential that the inner being is somehow part of a larger being as the thought of being alone is rejected.

To understand how a 'mind' can emerge from a collection of biological hardware it is possible to 'sense' the 'mind' at will.

Yes, that sounds like enlightenment in search of the 'power' but it is simply identifying an emergent Qualia.

The way to do that is to first identify which sense, aural or visual is dominant in your short-term, 'conscious' memory.

That is accomplished by (while reading this section) come out of the page, away from the words, see yourself looking out at the page and seeing the words and hear yourself reading the words in your head then select a few words and see them in your head instead of on the page and read them out loud from the image in your head instead of in your monitor.

So while you assimilate those instructions read on:

If you 'come out' of the page you back away from awareness of the page.

It is the same process that happens when someone interrupts your intent concentration in a motion picture or your 'being inside' of a book or your 'feeling' the music.

The deeper one enters such condition of 'becoming' the film or the book or the music the more an observer would brand the action as a 'trance'.

The 'trance' condition is really a focused and interested concentration of short-term processing which becomes far greater in the value of 'now' than long-term values of 'then'. Since a movie or a book or a song does the progression of logical steps for you the short-term process can simply accept input and send it to long-term which it remains focused as it is compared to the same input creating new short-term processes which sets up a loop of both short and long-term.

To be 'away' from the words is to read a word by its syllables instead of the image of the word as a whole.

Parse the word while reading it.

Long-term processing does not discriminate between parts, as it is not aware that parts are in a whole.

But your short-term process can discriminate parts from a whole and if you read words by their parts you will be able to deliver that speech without stumbling and if your voice is 'good enough' you may qualify to become an announcer if you don't mind the time spans between announcements.

To "see yourself looking out at the page" it helps to wear glasses as the frames give a reference point close enough to you to be considered part of you in your perspective.

As all wearers of eye enhancements know, one is essentially 'un-aware' of wearing glasses and becomes 'aware' of the glasses when they get in the way or when they are physically bothersome or when first placed 'on'. To "look out" is to be aware that you are inside, behind the glasses not somewhere over on the desk or perhaps in your left elbow.

To "see the words" one needs to realize that words are symbols. Symbols convey meaning they are not 'the' meaning.

Language is a symbol in sound where words are the symbols of those sounds to create the same sounds and thereby communicate the language.

When reading, a reader oft' times will read at a speed comfortable in acquiring the meaning of the sentences in relation to the next and previous sentence.

The words stand as parts of the sentence and meaning is derived by the concept created in acquiring the order of the sentences and phrases.

The single most important element added to the reading of any word or collection of words is 'perspective'.

If one reads a letter as the words are ordered and evaluates the concepts put forth by the symbols with past concepts observed by the reader the reader is totally ignoring and essentially rejecting the perspective of the writer.

When the writer's perspective is considered instead of the reader the purpose and meaning of the letter is understood as the writer intended it to be. Most people read a letter and come out of it wondering how it relates to their memories, if it influences or challenges any concept accepted as truth and fail to consider that unless the concept is the reader the writer did not include the reader in any concept.

To "hear yourself reading the words in your head" one has to first be able to hear words in one's head.

You can, whether you do, or you do not, already.

If you are an aural conscious thinker you will perceive words in your head and some may say they can 'hear' those words but the 'sound' is not 'a disturbance in a medium (a wave) it is an inner 'voice', although not too many will admit to a voice unless they perceive more than one and then they could lose control of both and become either.

Visual conscious thinkers may perceive words as well but not as 'sounds' in their head, more like 'presences'.

A visual person can do a wonderful favor to their brain by reading words out loud when reading text and listening inside to hear the word as well. It is a good goal.

To "see them in your head instead of on the page" one only has to look out at them to realize they are not inside and that the meaning they convey was intended to make a conceptual point and trying to reason what that point might be by accessing any previous point taken in elsewhere is most likely going to convolute the meaning intended by the author.

When "reading them out loud from the image in your head instead of in your monitor" can be a slow process at first it only requires being aware of each word as one reads it. That is accomplished by parsing the words as one reads them.

A word like "fish" is a single syllable and its singularity of being can convey all sorts of intended meaning.

It would depend upon the perspective of the author where the words "I was fishing for hope" meant searching (an implied process) or attempting to hook (an implied success) hope.

And if one completely ignores the author's perspective one might start displaying a Large Mouth Bass as a symbol of what was intended to be the concept of searching.

If you look at the arm attached to the shoulder attached to

the neck attached to the head that is reading these words (the left one) then you become 'aware' of that arm (or tilt wondering what happened to your right head) when at first you were not sure which arm so neither or perhaps both were 'known' or 'aware(d)-of'.

If you had problems with this paragraph read the one above it again.

If asked what that extremity is one may reply 'an arm' if one has little awareness of self or may reply 'my arm' if one has an awareness of self they can place in space as within their confined space which is not that arm but is connected to that arm and in charge of that arm (sometimes).

Reaching a level of awareness of any subject, topic, event, item, thing or concept requires simply winning it.

To win is to be any amount greater than the opposition or less than the opposition in some human games.

If nature's method were less than competition the victor would run out of food and evolution would have stopped at the concept of evolution.

A 'win' is a non-zero value. A lose is 0.

It does not matter if your team scores 72 more points than the competition, 1 will do.

That '1' is the singularity of that pathway representing that subject, topic, event, item, thing or concept.

The first singularity attempted by a growing brain is its balance of primary senses. Hearing and vision are those primary senses.

You do not evaluate your past based upon a taste or smell you evaluate your past based upon a reference to a taste or smell or some other less intensive process.

Vision and hearing are also the only two senses in the body that create a specific system focus center point.

In "*Meditations on First Philosophy*" [51] by René Descartes, born in La Haye, Indre-et-Loire, France, a lawyer by schooling who never practiced law who wound up in the service of the Netherlands military for nearly 20 years contemplated the "I" or 'self'. He sought a set of principals that were fundemental and realized that if an idea could be doubted it was false and that such a thing was deceiving and that if he was being deceived he surely must exist. It came to be known by a quotation that is not contained in the "*Meditations on First Philosophy*" : "*cogito ergo sum,* ("I think, therefore I am")." [51]

Since he was certain that he existed what form did he exist in?

Since he had discarded his senses as being trustworthy (after all they were the things giving him the feeling of being deceived) and he managed to overcome the deceipt through use of 'judgement' he concluded that the form he existed in was a 'thinking thing'.

The science that followed gave seed to calculus by the use of a coordinate system called cartesian (named after Rene') of intersecting points on a grid from a central 'origin' point.

This effort at merging Euclidean geometry and algebra became a rather precise method for locating where something is based on the system. [52]

A Cartesian coordinate system is close to the method of finding one's 'self' but lacks the ability to accurately define weight of contributing factors making a placement in a coordinate system truly imprecise as it ignores the perspective of the thing being measured.

Hearing is a direct Cartesian plot: but where Cartesian coordinates fail is in determining the location of an event not caused by the "origin" [0] point (center of the grid). That would be most things.

With the left ear providing a hypothetical value of 5 and the right ear providing a hypothetical value of 5 , both on the x axis the resulting position of the focal point of hearing would be the origin point. But ears do not provide totally equal values. Perspective alone tilts hearing values to the dominant side giving depth to sound.

Vision also places two values on the same x axis. They are simply closer together than hearing's value sources and perspective alone in vision off-sets the center point to give depth of field.

Rendering the Cartesian coordinate as a location of a singularity is replacing 'origin' with 'result' and default from 0 to 1.

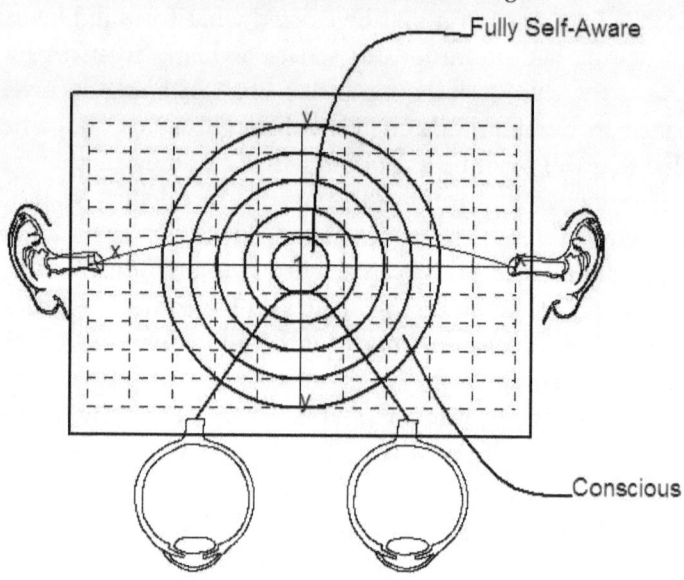

The physical confines of the 'system' determine the outer boundaries while the focal point of the 'system' defines the singularity or 1 value of equality.

When 'looking out' and absorbing instead of taking in and collecting the sense pathway that is dominant in you will determine the manner of which the perspective is generated.

Should your visual processing have reached the conscious stage of approaching singularity (within the outer circle) then your age would normally be between 3 and 8 years while the initial set in of non-zero value occurs just before that time frame (and gives rise to the 'terrible twos': a period when sensation is being replaced with awareness of sensation resulting in miscalculations until consequence is an awareness too.

By coming out of the page, away from the words, seeing yourself looking out at the page and seeing the words and hearing yourself read the words in your head and selecting a few words and seeing them in your head instead of on the page and reading them out loud from the image in your head instead of in your monitor acts to force the awareness of the focused sense closer to 1 or equality.

Another way to arrive at the sensation of awareness requires a friend or spouse.

Sit across from each other and simply converse.

While listening to your friend or spouse are you looking out at them and hearing their words in your head or are you taking in their words from outside and focused on them?

While talking to your spouse or friend take control of your visual short-term process by looking out at them, know that you are not outside of your brain and that the person you are talking to is not you and that you are the you that can be perceived to be in your head.

How well you execute that will determine how well you perceive a location of 'you' in your head.

But for now, read these words again and this time, do the words as well:

Come **out** of the **page, away** from the words, **see** yourself looking **out** at the page and **seeing** the words and **hear** yourself reading the words in **your** head then select a few words and **see** them in **your** head instead of on the page and **read** them out loud from the image in **your** head instead of in your monitor.

Did you do it? Only you will know.

Descarte was right about not trusting his senses because senses to Descarte meant long-term memory. He was also right in trusting his 'judgement' as that was short-term memory. How the two become one is the question of binding not the definition of existing.

 The two primary senses, aural and visual are split among many input receptors where each input receptor is pulsed to 'fire' twice each second. If they were all at the same pulse we would see life as a series of still and two dimensional images. Instead, we see life as a fluid and three dimensional existence as pulse rates for receptors are staggered across the field of reception, each firing within its pulse point two times per second. This blended input reception starts the binding process.

Synchronous firing of retina showing a firing rate (clock rate) of 1 per second (animated gifs can not duplicate 2 per second with this many receptors showing, actual firing is two times per second for each receptor) with 9 pulses per SP circuit.

 (Note: this is a reference drawing only and does not equal an exact count of receptors to firing rate. It does contain an image area of 97 receptors arranged in a staggered spiral method with each 9 pulse episode likewise staggered across the face of the retina. Since each receptor (using the lens method of a single double convex lens either on a concave surface or an optical adapted flat surface for concave presence *See Right) is a quasi-holographic representation of the complete image viewed.

The only difference is perspective of its location on the retina and its firing episode. Retina is depicted in non-linear image representation. [53]

Each input pulse represents a control valve (or regulated

70

resistor) that permits an amplitude pulse of the subject's base wave frequency to pass into the processing chain in the order it was received by that receptor.

This serves to convert external stimuli of any form (all senses) into the single processing method employed by the brain: the second level of binding.

Your computer providing the information the monitor is displaying to you is capable of sampling an external sensor twice each second and is capable of converting the measurement used to sample into a representational number.

Your computer can then perform a task based on a predetermined number, range of numbers or prior number (if it is recording its previous comparisons).

That would be 1:1 calculation.

Input equals output and if input is fast enough then output will appear just as fast and if both are near to equal to the thing being measured a relatively believable representation of that thing can be generated virtually. That is essentially the science of Artificial Intelligence. It is using a machine process to replicate not duplicate.

The brain employs 'exponential processing'.

It has to.

With speeds like biology can provide naturally and react to fast enough the range of processing speeds must remain very low.

That requires the 'system' to be above all else, efficient.

Exponential Processing is the system that runs your mechanical watch.

It is the system that gives your automatic transmission the ease of shifting gears without damage to the system. It is the concept of the gear.

In this analogy the input receptor 'gear' has two teeth in opposition. Each full circle spin of a one second base frequency turns two revolutions of the input receptor's gear stepping up the processing from 1:1 to 2:1. (In the past I have referred to this process as 'ratio enhancement' but have since dropped that term as it describes the result, not the cause).

This scaled (12 teeth instead of 30) gear represents the speed of processing of long-term

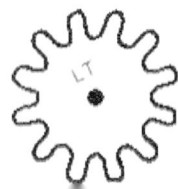

memory.

With each 2 times per second rotation of input's speed gear the long-term memory gear moves thirty teeth.

A second gear is likewise attached to the center pin of the long-term gear that causes 900 teeth of the smaller gear to pass for each single tooth movement by the larger gear.

This stepped 'up' process is exponential processing.

In biology is it accomplished by division.

If the base frequency was 35721 hz the first input receptor's pulse rate would start a firing at 1hz and at 17860.5 hz.

Each additional input receptor of the same input source (eye or ear), each either tuned to a specific reaction frequency (rods, cones) or (fenestra ovalis conducting fluid waves in frequencies to hair cells associated with specific frequencies) is pulsed in a wave action..

Long-term memory pulses would occur every 595.35 hz while short-term memory pulses would occur every 19.845 hz. Slower rates mean more processing and therefore efficiency becomes a factor.

When 60 hz is compared to 1800 hz by each being stepped down to 10hz the stronger source (highest amplitudes) will be the controlling source and you will be either visually aware or aurally aware in short-term memory.

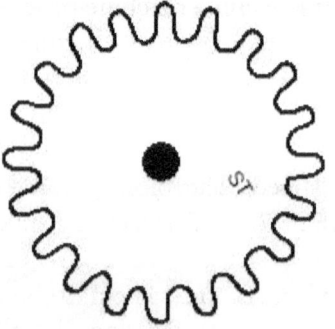

It will take a few years for your long-term memory to pick up on that steady range of amplitude values and as it assimilates them your short-term may question the presence of a foreign entity and call it a name and make it an imaginary friend or a monster under your bed.

As you search for perspective the dominate processing pathway will guide the evaluation.

Brains other than human (with some slight similarities in quadrapedals) do not have the luxury of perspective.

Animals are degrees of good and bad, positive and negative in everything we do, regardless of species.

Only the human brain can reason alternatives and override the defensive default to accomplish a concept.

Where we might declare we would really love to have one of

those chocolate cupcakes and drool over the thought of its smooth and silky body and moist deep chocolate icing our little friend the family dog can only recognize the smell as a degree of good and the sight as a degree of good. Good might entail extra tail wagging but it is still a degree of good.

Being simply 'other-aware' the over-grown puppy is not aware of the manner in which we treat her like another child she only knows it to be a great degree of good.

Our power of self-aware perspective interprets the body actions, sounds and especially the facial and eye expressions as being emotional reasoning until we realize that our subconscious process is also emotional and if we do not care to use the other one, that short term shorted out loop of self-aware memory trickling down the long term tube then we are only different from animals by the ability to say "I".

If we let subconscious processing rule our lives we will have a ruler we have already suffered through over and over again.

Tyrants and con-men alike keep control through the support of this subconscious process by playing up to its weaknesses and fulfilling its desires.

Animals simply know something is or is not, not what something is. "Good." Will suffice for our baby since her default condition is good.

We have experienced the other default condition before and know how to steer clear of that herd.

That, that, see that, that. Good. That GOOD. THAT GOOD. THAT GOOD! THAT GOO…

Until the wagging and the panting and the shuffling of claws on the door finally reaches the point where my highly developed self-awareness blurts out "she really must want to come in".

She really knows that door when opened, is good.

Talk dog sometime to yourself.

Do it all day long.

Don't use "I" or "me" or even possessives at all.

Simply 'that' for every target of attention and targets of attention must be addressed unless they have been addressed before.

The more addressed the less addressing is required for it to evolve into the background.

Take a day and walk the house only responding to noises you

hear or movement you feel or things you see that jump out of the background because they are different by some degree.

If you are a cat person and care to take on the feline it all still applies.

You only have to change your perspective from the aural of dogs to the visual of cats.

Take a day and walk the house only responding to things you see or things you have seen.

The more you see something the less you react to it.

The older you get the more it takes to be reacted to.

Trust you'll spend most of the day avoiding such controlling environment and sleep on the couch or perhaps curled up under the bed or finding a nice sunny windowsill to sleep off the aggravation of it all.

Singularity actually means "the quality or state of being singular" [MW] which the quantity of zero cannot fulfill as zero is nothing so it cannot be singular. The only value a singular may ever have is a non-zero value.

We experience a non-zero value in our being conscious. (Not the same as self-aware). It is that non-zero value that sets up the 'mind'.

The 'mind', being an emergent property of the architecture of a closed loop memory system is self defined, self-imagined and self thought through external stimuli.

The degree to which the 'mind' or short-term memory process is in control of the result of the mixture of long-term memory output and short-term memory output determines the observable intelligence of the subject.

While Binet's "g" or intelligence factor is tested using standard IQ tests the test only works on long-term memory.

It does not provide nor consider any indication of short-term processing which is where human intellect resides.

The speed of the short-term memory's processing amplitudes in relation to the input stimuli processed amplitudes; a result of long-term speeds determines the available intelligence of the subject.

Brains other than human have available intelligence alone.

Human and other forms of brain 'thought' are wave amplitude variations which through architectural blending of identical frequencies in variable amplitudes within neuron chambers, pulsed in the same synchronous pattern, and returned for comparison in the same 'time-frame' as the input (unsupported means 'smaller' amplitude therefore older memory) results in the emergence of observable intellect, as long as the process flow remains in the single direction of its architecture which is controlled by the one-way flowing synapse, the biological diode.

Emerged collection of individual pathways from the same focal point provide not only focus of the whole in 3 dimensions through partial out of focus peripheral receptors, but when architecturally combined to form a central focal point results in the perception that a 'mind' exists within the head, the central focal point of hearing and vision.

Being the only location in the architecture that is dual focused from stereo-optic perspective of each, the head, which luckily also holds the brain responsible for this binding is the focal point of the two most active input receptor types in short-term memory processing since the pattern established by the architecture in humans results in a dual closed loop system where the outer pathway is down and the inner pathway is up, the outer pathway performing all memory amplitude 'calculations', causing reduced amplitude with each neuron event setting up the concept of 'older' or 'time'.

This identical process in long-term memory is running so slow that an emerging self-awareness is a very long process and elusive or a very fast process if nearly all focused senses find an equality through input stimulus.

The same focal point of inputs provides the equality in all brains but the awareness of it is evident only in closed loop short-term memory processing humans.

Short-term processing is running at a power of two over long-term resulting in setting up the closed loop of a single existence, where the comparison of pulsed variable amplitudes is equal with the value presented to it from long-term processing making the short-term not have a concept of 'time'.

Time is considered to be the measurement defined as "the period during which an action, process, or condition exists or continues" (MW) (which would be the definition of the measurement) or as a 'non-spatial continuum' (a coherent whole that does not displace space) .

Time as a measurement is definable any way a society deems fit.

Time as a continuum, spatial or non-spatial is a result of amplitude variations in the brain and stops at one of two junctures:

1: When it no longer can be evident:

(emergent whole has become equal, self-aware) or

2: When it no longer can be evident:

(subject is deceased).

Since humans enjoy both long-term and short-term memory processes and the short-term memory process alone will emerge as equal after 3-8 years of use and the awareness of things a brain does is from that short-term memory process's closed loop the process doing all human 'conscious' output is both a product of long-term memory and the cause of long-term memory.

Both memory levels operate independently of the other through an orchestrated symphony of processing patterns, sharing neuron chambers for execution and at the same time as a single entity as long-term feeds short-term from external inputs compared to long-term's depth of recall which processed in short-term feeds long-term resulting in the 'memory' of being self-aware.

Both levels of memory processing output to motion and compete with each other for dominance over motion.

When motion occurs and effects the facial muscles without prior conscious or short-term command control the event is viewed by short-term as being foreign and therefore not 'understood' or 'recognized'.

Those outputs to motion occur in all muscle groups, chemical triggers and input receptor types and are either controlled by long-term memory process or short-term memory process or a degree of both.

When nearly all focused senses reach equality through input stimulus the focal point of that stimulus can be said to be 'loved'.
It is essentially distributed pathways of varying sense types becoming 'aware' of a connection to another entity.

Had history recorded the event of the first utterance of a spoken human word it might very well have recorded the overwhelming desire of a female to aurally express her visually based feeling of equality with another human being.

Had a male been responsible for the first spoken word his long term aural dominance having left a weaker visual long-term memory would seek to satisfy the difference through visual stimulation and could very well have given the first thorny flower to which the first word spoken in response could very well have meant, "ouch", in a new language.

Emotions are 'emotions' because they are both internal and external and like the 'mind' seem to reside in the head but seem to also 'take control' of the body to a degree or another without the 'mind's' awareness.

An emotion is an unaware execution of reactive response that takes place in the long-term memory output process to motion.

With long-term in control over motion it is also in control over internal-output (the variable amplitude presented to the closed loop of short-term memory processing) resulting in an 'awareness' of a degree of difference between the 'normal' with no emotion and the 'emotional' through the presence of it.

Every "emotion' is to some degree a degree of the preponderance of either good or bad, positive or negative, true or untrue, of extremes.

The degree the 'emotion' varies from 'normal' amplitude ranges in short-term processing establishes the degree it is 'recognized' or 'identified' by short-term processing.

Low amplitudes cause the short-term 'mind' to perceive a slow condition. High amplitudes cause the short-term 'mind' to perceive a fast condition.

How that condition is executed in the combination of supported long-term memory and how that condition is displayed in motion determines 'what' emotion 'name' it will be given to the behavior displayed.

An emotion is a result of long-term memory processing, not known by short-term processing, slowly changing over 'time' to potentially dominate short-term processing and thereby remove short-term control from motion altogether.

It is a process that is architecturally based, therefore very 'machine-like' and results in a collection of timed outputs from a centrally perceived 'mind' within a focal point of reference.

It is our sense of time that gives rise to the need for order and our need for order that gives rise to the notion that time somehow exists outside of what gives its awareness to us.

Luckily for us the two main processing pathways of the brain are both located in the same close proximity area and are accessible by addressing them directly or indirectly.

After the answers to the Top 10 Questions About The Brain (in all 24 varieties of tied poll questions) a simple household amusement will demonstrate a wave within a wave and how parts of a whole are important while the sum of the parts is no more important than its smallest part.

THE ANSWERS:

What is wrong in the brains of people who suffer from Schizophrenia? [5]

If the condition is genetically traceable the problem lies in equal clock rates among distributed input receptor pathways of the same focal point or sense which would result in lower levels of dopamine in memory feedback loop pathways.

If the condition is not evident in the family tree of the subject the condition is most likely attributable to the other cause of lower levels of dopamine in memory feedback loop pathways.

Short-term memory concentration has focused on either a standard or repetitive event or concept, which has connections to parts of prior events but is not directly connected (a previous comparison has become its own comparison).

Schizophrenia is essentially defined as the "lack of interaction between thought processes and perception." [5]

The 'thought process' is the action while the reaction is the recall or awareness of a place in order. Regardless of psychiatry's insistence in positive and negative forms of schizophrenia there is nothing positive about being out of 'synch'.

What 'is' out of synch in schizophrenia is the opposing sense. If the opposing sense is the dominant sense what goes in will not come out with reference to its having come in as the same thing therefore perception will not equal intention.

It can be solved by reinforcing the weaker sense through acquisition of a sense of true self awareness in that sense.

Such increase amplitudes will require higher levels of dopamine to transmit across synapses and the evidence of schizophrenia will disappear.

If one is aware of one's self one cannot be controlled, nor influenced by the awareness of anything or anyone else.

We all suffer a degree of schizophrenia far lower than the abnormal results required for diagnosis when we do not have a central command, a leader in charge.

When we do not have full self-awareness we are victims to what we were victims to before and with each new victimization a reinforced memory of past victimizations remains strong in recall.

Do we be more happy, if we solve all the unknowns of brain?

That depends on your definition of happy.

If by 'happy' you mean, "enjoying or characterized by well-being and contentment" [MW] it can only be so if there is not too much prior subjective deduction standing in the way.

If it is, and if the short-term doing the evaluation is not completely in control over the long-term's rejections of unknown concepts you will prefer remembering Steno's assertion and feel comfortable that you are not alone in your rejection of an unknown thereby excusing your long-term from its concerns for longevity by rejecting the short-term's desire to find the parts and make connections there; so perhaps your recallable, supported and most often trained long-term memory will win retaining the non-zero control over the 'mind' through recall of past recalls of past recalls.

If by happy you mean, "favored" or "enthusiastic about something to the point of obsession"[MW] then by all means, yes.

If you knew how you thought and you knew that you knew how you thought then you could think about knowing that you know that you think about knowing.

That alone can take a lifetime if you focus on one subject. Applying it to everyday problems means tackling instead of placating.

It also means creating artificial machines to do the same process can become an obsession with specific purpose instead of an obsession for a purpose.

Why do psychiatrists tell us what to do when they don't know

how it works? [6]

Psychiatry is rooted in observation and trial planted in those persons with a desire to solve problems of mental causes either for their own gratification or for the gratification of seeing others grateful.

It is amazing to note although that the term 'cause' can be so eloquently pigeonholed. Once a 'cause' under this confined definition of the word is identified it becomes removed from psychiatry and falls under medical science.

"Historically only two psychiatric disorders have been definitively linked to causes. General paresis of the insane (now known to be a late stage of the disease of syphilis) and pellagrous insanity is caused by niacin deficiency (the mental result of the disorder once referred to as scurrvy.)

Ironically neither of these are any longer thought of as psychiatric disorders." [6] That makes every single current mental condition treated by psychiatrists to not have known causes. Since nothing is defined just about any treatment if believed by the subject has the same chance at working.

Psychiatrists find success in one form of therapy or another and continue to treat results of brain conditions with drugs that mask the result without addressing the cause and therapies that support the subjective instead of unmask the objective.

Instead of the frequency of success being by addressing the objective cause it is the frequency of success by addressing the subjective result which does, to some degree, reduce support connections to past problems but in no way increases the connections to current relevance.

I suppose such 'find the cause but if you find the cause we can't treat it any longer' mentality justifies keeping therapy sessions running into the years in frequency visits where the same subject having the same results from a repair person who comes in once a week to remove their TV for an hour to fix it and years later continues to insist he can still fix the TV would sue.

Now' is not what we experienced: it is what we do with that experience whether we blindly follow a leader into becoming the drone needed to remain a follower or we choose to take control ourselves and lead.

What is 'the mind'? [7] [48]

The singularity of the short-term processing pathways of the human brain.

What are dreams and nightmares? [8] [9]

When you sleep the only difference in your brain is the level of input amplitudes passed for long-term processing.

When long term's 'conscious' non-zero value falls to '0' one half of the output to motion (short-term's contribution) is shut down.

Once that occurs you enter a period of 'light' sleep which is long-term memory fed by nearly no amplitude value.

As the amplitude value decreases one may experience a half awake – half sleep condition and 'dream' or be 'aware' of a 'dream' that is of more recent memories.

For a period of time nearing 90-minute cycles the values of long-term memory are reached that are equal to the near greatly reduced input values.

When that occurs we enter what has been termed 'rapid eye movement' or REM. The eye movements are the only motion of the body (normally) at that period as the eye is the input receptor for major muscle groups and motion to it is coming from the reactive output from long-term processing.

As a series of near equal values filters into short-term memory it is passed back into long term memory near the same level not supporting the very old memories being accessed just replaying them.

Since old memories are accessed without the benefit of the 'now' time reference provided by a 'conscious' awake 'mind' many comparisons are made of similar values whether they are in the same time frame or not.

When the period of low values is met by a wave amplitude increase of a previous a period representing a previous 'awake' state (returns in feedback are not 1:1 they are spread out over 1 so as to be more similar to 1:6 the 'time' reference is 'compressed' and the result is an experience not relating to reality of input 'now'.

We only become 'aware' of a 'dream' when we are in the wake up period of increasing amplitudes are input receptors start to permit amplitude variations and it is then that we can discern and mark the 'dream' as a 'dream' or a 'nightmare'.

Some 'nightmares' invoke much strong outputs from long-term to motion to cause a 'friendly' experience, a sleepwalk event or a 'bad' experience we would then refer to as the 'nightmare' that woke us up.

Regardless of how it is accomplished a 'dream' whether good or bad is only a figment of past figments or past memories not related at all that become related in amplitude values so low that they do not register in recall during the 'awake' state. The discussion of 'animal' dreams in question 16 would be of interest.

Explain: answers to the "hard Problems" of consciousness study (brain mind binding, explaining subjective experience and causes of consciousness, will) [10] [11] [12] [13]

Just about all of these issues were addressed in the opening and Brain 101 sections of this chapter.

What was not addressed was 'will' which I will not assume was meant to imply that is without cost in some form or another.

'Free-Will' has taken on a portable meaning which undefined things have a habit of doing.

'Will' of any manner is an implied action based upon a choice.

A choice is not a choice if any of its options is contrived. A contrived option removes 'will' and replaces it with 'reaction'. The only method to acquire 'will' is to have a part of a system take control over another part and impart its 'will' or method over the 'will' or method of the subservient ingredient.

What is consciousness? [12] [15]

Consciousness is the state of being aware enough to know of being aware at all. It is a non-zero value over not being aware.

The sense of 'self' or true awareness is not achieved at a non-zero value it must have amplitude over other amplitudes for it to stand out and it is not the normal default state of a human being.

The normal default state of a human brain is the non-zero value of consciousness. Allowing that non-zero consciousness to control your life amounts to allowing the past (long-term memories) to slide right through short-term processing setting up a default 'on' condition in long-term that rules your life from past rules.

That is the goal of intensive training.

That is the method employed in education; the method tested by Binet IQ tests and the reason for susceptibility to depression and other internally supported non-aligned with reality conditions.

The default condition of non-zero consciousness is so prevalent that those few who do use short-term for more than a way station are relegated to mistrust and disparagement by their refusal to follow accepted practices, write in a manner in which ignorance is supported or follow the same leader.

Short-term memory can be employed by all humans to override the default conditions of reactive long-term processes.

Is it true that we only use 10% of our brains? [14]

No.

Not at any time.

Any brain cell not used is lost.

There is a wave of the base frequency pulsed in clock rates equal to the division of the pathway's input receptor present in every living neuron.

Without the non-zero value of the frequency (no amplitude) there is no life to make the neuron 'alive'.

If "we only use" refers to taking advantage of: the answer would still be no. All parts of the brain are doing their 'thing' at all times. During sleep all of the brain functions except the input values affecting the initial amplitude are reduced providing very small amplitude with each pulse.

Can you exercise your brain? [16]

I do it all the time, thanks for asking.

Oh. Yes.

But be careful what part you exercise.

If you exercise long-term memory it will begin to take control and you will follow what ever took you there.

It happens in short spurts with every conversation.

A topic is started that relates to something it is made of which relates to something it is used for which relates to something the use of it has to do with how it was last used which relates to the results of how it was last used and what the topic was is no longer the topic and the conversation is 'interesting' but worthless.

Exercise your short-term memory by playing a game of cards being aware of the game and the rules at the same time.

Exercise your short-term weaker sense by forcing it to meet the dominant sense:

The exercise provided earlier in this piece for gaining the sense of self is one very good way to use idle brain time but should be practiced in periods where brain power is necessary and it will become more and more normal until you will be just as weird as the those great thinkers of the past but you will be thinking 'now' and learning from the past instead of living it over again.

How complex is the human brain? [17] [19]

From a biological perspective it is complex due to its quantity of connections and components.

From a system perspective it is very simple but nothing at all like the perceptions placed upon it by subjective reasoning.

What is wrong with Rodney Brooks' viewpoint on the brain? [20] [41] [42] [43] [44] [45]

Professor Brooks has a brilliant long-term memory process.

A normal long-term memory process would make the amplitudes of long-term memory in the same range as the amplitude of that level's base frequency.

An advanced long-term memory would return more amplitude than the base frequency and a brilliant long-term process would return nearly the same amount of pulsed amplitudes as are being created.

If that person was a visual long-term thinker (male, or the rare female) their 'photographic' recall would be celebrated or the cause of their bouts of depression would be consuming.Just as the long-term rate of amplitude return for processing is able to be in excess of 'normal' it is also able to be less than 'normal'.

Less than normal long-term recall can be observed in habits, and learning deficiencies and the lost touch with a sense of 'now'. Professor Brooks undoubtedly does not suffer from any long-term reductions.

And given the depth of his reasoning does not suffer from a short-term reduction.

He does though, have a focus: robotics, the replication of the living.

"In the 1970s both the USA and UK suffered funding contractions in AI research, with specific criticism of AI for ``playing around'' in ``toy worlds'', which did not have ``industrial relevance''. This put a damper on such complete systems --- which could only be constructed in simplified or ``toy'' worlds --- and caused a shift of research emphasis towards implementing systems which contained all of the complexity of real industrial problems.

Of course, *complete* systems with industrial-strength complexity could not yet be built, so this insistence on ``industrial relevance'' merely changed the kind of simplification researchers employed: from implementing complete systems in simplified worlds, they shifted to implementing realistically complex fragments of complete systems. ...

This change constituted a major paradigm shift. In order to accommodate this change of experimental method, roboticists had to switch from a research program underpinned by the ``sea-slug'' assumptions, to one underpinned by the assumptions of classical AI. This switch was facilitated by the fact that the presumptions underlying their approach had not at that time been made explicit. In other words, I suggest that the switch in experimental methodology consequent on the criticisms and funding contractions of the 1970s entailed a switch of underlying presumptions of which the researchers of the time were not explicitly aware. This illustrates the danger of pursuing research without making one's basic presumptions and philosophy explicit. " [20]

"Note that Brooks was at this time already a major

contributor to the classical approach in assembly robotics. It was while collaborating with Lozano-Perez on a paper defining the next major phase of MIT assembly robotics research that he decided that the whole approach was irretrievably damned, and switched to experimenting with behavior-based architectures in mobile robots. As a consequence, what had been planned as a major publication leading a grand research program into the future became a speculative paper overtaken by events." [20]

When the method was behavior modeling the behavior is a subjective result and it leads to attempts to create a brain by 'building' one like evolution, program one like chat-bot builders or build one and let it emerge into intelligence and consciousness and self-awareness: never mind the insignificant potential of Cog ever becoming 'conscious' and realizing the center of consciousness was next to him.

In the Professor's own words: "Building an android, an autonomous robot with humanoid form and humanlike abilities, has been both a recurring theme in science fiction and a "Holy Grail" for the Artificial Intelligence community." [42]

With all due respect, the Professor is missing the point.

His justification of behavior-based modeling has become his 'holy grail' when the 'holy grail' of Artificial Intelligence is described elsewhere as "The holy grail of AI is to create machines that can truly mimic the human brain in the way it thinks, responds and interacts. " [55]

The way it thinks, responds and interacts has no relation to what it does other than being its cause.

The Professor has settled for results as able to be replicated and accepted causes as unattainable then believes that somehow some emerging system will manage to be programmed that causes the Cog device to not only look human but become human-like in knowing it is a thing unlike any other thing.

There is no memory in Cog that can relate to any other memory in Cog without an instruction.

Therefore Cog is a toy making MIT's contribution to the science of understanding the brain enough to duplicate instead of mimic about moot.

"Our alternative methodology is based on evidence from cognitive science and neuroscience which focus on four alternative attributes which we believe are critical attributes of human intelligence: developmental organization, social interaction, embodiment and physical coupling, and multi-modal integration." [55]

All of which are results like apples not the tree.

How do neurons carry information? [18] [46] [47]

Neurons 'carry' information by being the chamber where interaction of amplitudes takes place.

There is a corresponding cellular action, which represents the non-zero value of the amplitude frequency being processed in the neuron which gives rise to the 'firing' or binary state of a neuron's condition.

While the 'charge' can be measured it is indicative of a process present not the process itself. The process itself takes place in the wave opposite in charge to the cellular frequency. This process was attempted by fuzzy logic.

"Fuzzy logic is a generalization of standard logic, in which a concept can possess a degree of truth anywhere between 0.0 and 1.0. Standard logic applies only to concepts that are completely true (having degree of truth 1.0) or completely false (having degree of truth 0.0). Fuzzy logic is supposed to be used for reasoning about inherently vague concepts, such as 'tallness.'" [54]

Where the wave is amplitude variation from the near non-zero value to the 'tallness' of the base frequency's maximum amplitude fuzzy logic tries to replicate that system as a degree of non-zero as well but models 'subjectives' instead of 'objectives'.

Why have we got a memory and how does it work? [21]

Imagine yourself without a memory.

You could not as the moment you imagined you would no longer imagine.

Memory, as used in computers is a specific place, a specific value and a specific order.

Memory as used in the brain is a specific place, a specific value and a specific order in order of pulse, in massive parallel processes blended through input firing order rates and internally provided an output that is the sum of the parts.

Memory works the same way brain calculations work.

Each memory 'cell' is a neuron that processes a pulsed amplitude wavelet by comparing it to the base rate of that level of processing.

The result is a wave within a wave.

Wave amplitudes grow as they first fall into memory and then fall as they get older, this provides the sense of 'now' in relation to a moment ago and keeps out being able to discern 'now' as anything other than fleeting and sets up the relation to the time order of processing in which to compare with previous memory to form new memory.

Every memory you have is a relatively new memory. Unlike computers that place a value on a coordinate location of a hard drive disk brain memory goes 'down' the line where it is copied and sent back 'up' the feedback line.

As it travels from one neuron to the next in sequenced order back to processing it is processed with the base rate again of that level of processing which reverses the wave set up going 'down' memory and returns the memory in the same relative relationship that it went in with. How can a there be a wave within a wave?

See the Uno® example at the end of this piece and earlier in chapter two.

Are boys smarter then girls? [24]

Males and females are indeed different but smarter is not the measurement.

How smart a person is has to do with that person, not their gender, race or personal desires. Human females are primarily visual long-term creatures.

Human males are primarily aural long-term creatures.

How smart a person is, is determined by how much the short-term is in control over long-term in the battle to rule the body. And then 'smart' can be judged instead of 'knowledge recall'.

What will it take for an artificial brain to achieve philosophical consciousness? [22]

Whatever time you wish to proclaim it as long as consciousness is not defined.

Reaching the internal amplitude level within a single input process' ensemble of orchestrated pathways will take building it and perhaps 3 years of nurturing its growth and environmental perspective.

What does the brain do during sleep? [23]

The same thing it did while you were awake.

The only difference is while awake the system was being given higher variable amplitudes and in sleep it is being given much less amplitudes.

To a human being the process of sleep is a retreat from reality as well as a charging 'moment' or recharging event.

But to a creature not human (not given output from short-term to long-term but from long-term to long-term) sleep is nothing different than 'awake' other than being moments of different things.

There is no 'conscious' condition in a system without a circular connection. The upcoming (at the time of this writing) Discovery Channel 'exclusive' about animal consciousness is just another in a long stream of visually dominant scientists and film producers imparting their perspective on other creatures because they fail to comprehend their perspective is made from the manner in which they primarily think and is not at all like everyone else primarily thinks.

Yet we sit and watch the speculation presented as fact and it soaks in and supports Steno with every production.

Are there hard and easy problems about the brain? [25] [26]

Only if you want there to be:

If one wishes to explain their inability to explain something and the only hope of such explanation is that it explains the explaining and not what so far could not be explained then one could impart degrees of difficulty to the acquisition of answers likewise people adhere to.

The notion that some 'problem' with the brain's understanding can be judged in degrees requires the assumption of a priori. When something is self-evident, like Descartes' senses being untrustworthy or that moment of self-awareness most readers will have experienced after reading the above short-term acquisition exercise again, it is a thing that has become 'aware' to the brain.

A person can become aware to a brain and be loved or what that person represents to the brain becoming aware if based in a desire may be lust. Either way it is the same process.

The fact that it has become an aware emergence of brain processing does not mean it is correct or good or right or truth.

Does the body have a mind of its own? (or: why does neuronic signaling sometimes appear AFTER the act - see Binet et al.) [27] [28] [29] [30]

Yes, it does.

It is housed in a specific part of the body which occupies the most upper extremity also housing the eyes, ears, mouth, nose and sits upon a part of the body that gives it independent movement from the body.

The brain IS a part of the body.

It is not a special thing.

It is what is going on inside of it that is special.

The reference to Binet is indicative of the desire to justify the concept of Binet's assumption: "that lower IQ indicated the need for more teaching, not an inability to learn."

The IQ process developed through Binet's work measures long-term processing results and is best applied to long-term processing where a person with nearly no short-term control can remember logical order processes and comparsions of minute degrees through unimpeeded recall of long-term concepts.

A true IQ test would measure the degree to which short-term is relating to long-term. 'More teaching' has been construed as successful implantation of known and accepted 'state of the art' knowledge.

Without a single reference to how to use that knowledge in the short-term reasoning process a person is tested for their ability to recall concepts created through knowledge.

High IQ persons are more associated with 'different' or 'weird' than 'normal' or 'safe'.

How and why does the brain decide to block out memories of events, and how does it keep them blocked out (until/if retrieved)? [31] [32] [33]

The use of the term "block" implies "interruption of normal physiological function" [MW which is hunting for the haystack by studying the needle. Instead of implying an obstructive method consider the opposite, that of a reduction method.

If a person experiences a serious jolt to sensibilities, one that does not relate to past experience, does not fit the standard accepted by past experience, that jolt will not have any specific concept in memory with which to be supportive.

As a 'first' event memory the event may not be recallable unless it relates to a previous event. If the event is sexual in nature the jolt will disrupt the sexual concept without direct reference and may be experienced as a rejection of advances for an 'unknown' reason. Memory in the brain is not like a computer.

A computer places a bit value of 1 or 0 in a grid position on the disk and it is going to stay there as long as the disk works unless it is changed or rendered not used (erased).

(There are not 'three' possibilities to a computer memory, either 1 or 0).

The act of 'erasing' simply leaves no current file connection to a registry entry. It does not change the memory's state.

In the brain memory is fluid. It travels in order of timed pulse of input and gets larger and smaller and slightly larger and slightly smaller until it reverses and returns for comparison to new input.

Memory does not reside in a specific neuron or specific 'space' it moves through specific neurons and specific 'space'.

A dog with bad memories will live those bad memories over and over again when confronted with similar stimuli.

Humans are able to take control of those long-term

memories and impart a relevance of them to 'now' which reduces their influence on 'now' and lowers their recall-ability. Recalling them is still possible by allowing the depth of return to work on the same concept longer than long-term's in out process.

That requires short-term processing to retain the subject. When short-term processing imparts the relevance of an old memory to no relevance to 'now' that memory is not supported and falls smaller in amplitude.

Is the whole (consciousness, thought, awareness) greater than the sum of the parts (tissue, electrical signals, neurons)? [34] [35]

Not in the least bit. It simply is the sum of its parts.

The brain is part of a biological machine and is just one part of that machine. It is the part the essentially runs the machine but it can become dependent upon the whole machine and ignore its abilities and simply exists.

That does not make the whole of the brain, the collective result of a working brain any greater than the smallest amplitude pulse wavelet within it.

Does decentralized functioning exist..?[36]

Watch a flock or birds.

Observe a school of fish.

Alone, each creature is but a part of the whole as the parts of the whole move the parts of the whole maintain order and progression and follow the movement in near perfect harmony acting upon their position and reacting to the change of their reference.

The flock of birds or school of fish represent a decentralized function but do not represent the binding of such a decentralized function to become its own function and therefore its own being.

Each part of a school of fish or a flock of birds is a single thinking machine where environment alone determines reaction based on past environmental reactions and the 'thing' remains a subjective as it is not an objective singularity.

Each creature within the group perceives the environment based upon the method of thinking most dominant in that species and that specific creature. It is not a hard thing to comprehend that a thing can be made of its parts that are semi-independent (in that they are focused from the same perspective machine) and merge them to become a single entity that is not the same thing as its parts.

Are all brains born equal..? [37]

By all means but two.

1: components within the brain are defective or diseased

or

2: the frequency of the base wave of the brain is an even one. Unless a specific sense is unequal to its companion counterbalance sense (vision-hearing, taste-smell, pressure-temperature) the brain will be equal with human females slightly advanced in long-term visual memory and human males slightly advanced in long-term aural memory.

If the frequency of the base wave is even it will not remain coherent well within a closed and close proximity shared processing environment and will cause the recall of larger amplitudes to not be represented by their correct 'height'.

An even frequency broadcast throughout a closed space sets up interactions in timing that are not provided by the divisions of the bodily clock (which itself is based in the base frequency.

Consequences that should register as warning signs of repeating behavior are not comprehended as 'important'.

Why do some people have obsessive personalities and some don't? [38] [39]

Imagine a motorcycle where the wheels were not traveling at the same speed.

If the wheel that was lacking in speed was in the rear it could be said to be always trying to 'catch up' and if the wheel lacking in speed was in the front it could be said to be always trying to slow down the pesky rear wheel.

Should the rear wheel (which is perceived to always be the cause of the problem) represent long-term aural processing and the front wheel represent short-term visual processing (a visually dominant short-term memory) the visual short-term 'mind' would be faced with a stream of nearly continuous inability to connect a concept with a visual.

If the rear wheel (still at fault) represents long-term visual processing and the front wheel represents short-term aural processing (an aurally dominant short-term memory) the aural short-term 'mind' would be faced with the need to complete an image it is unable to create.

Need to fulfill a semblance of logic and order can be manifested in compulsion but normally compulsive people are compulsive because they are reacting to long-term memory outputs without much short-term control to balance the reaction with contemplation.

Why are there so many definitions of things no one knows about? [40]

Every person has their own perspective on observable norms.

What may be caused by one thing to one person may be caused by another thing to another person.

Most often the only thing in common with the perceptions is the thing being perceived. Therefore a name can be given that is agreeable to all perspectives and the more agreeable it is the more reinforced it will become and the most resistant to being tossed out it will be as the closer it reaches the point of becoming its own topic.

The more support of that new topic the person is subjected to, especially in teaching environments where retention of what was learned is paramount to the evaluation, dissection and repair of what is being learned the more the topic becomes real on its own accord.

The closer to the perception of real it is to the believer the more support it will receive from 'believers'.

Whether it is mythical notions of sperm constructed moist brains or the belief that one can ever lower one's goals to pretend to reach one at all the act of knowledge acquisition left unparsed, unevaluated and supported through arranged methods the result is the same: If evolution ever started on the path of the rules of golf we would have no Pebble Beach.

WAVE WITHIN A WAVE:

"The Elliott Wave Theory is named after Ralph Nelson Elliott. Inspired by the Dow Theory and by observations found throughout nature, Elliott concluded that the movement of the stock market could be predicted by observing and identifying a repetitive pattern of waves. In fact, Elliott believed that all of man's activities, not just the stock market, were influenced by these identifiable series of waves." [56]

Could a wave be present in a simple game of Uno®?

Chart one. Two people play a 5000 point game of Uno®. Enter the points, totals and round points in separate columns into a spreadsheet like Excel. Chart them. Look at the wave within a wave and how the entire game is running its own game with the players thinking they are making it happen.

Chapter Four
What Is Dyslexia?

The majority of participants of the question poll seemed to be individuals interested in Artificial Intelligence, which according to the moderator of the comp.ai usenet discussion group " It's about the brain ... so, comp.ai.philosophy or maybe sci.cognitive. Nothing for comp.ai here."

It is no wonder AI is anything but a series of guesses, proving each guess until the next guess gets favor and wasting investor money in the process.

Dyslexia is one of the results of a mismatched brain.

Input signals in the brain are first processed by long-term memory in which to give the short-term 'mind' something to think about. It is why you can do something and then become aware of it.

The same situation is at work in dyslexia and causes a word or letter or number viewed by long-term processing to be output to speech before short-term processing is aware of it and normally while short-term is still thinking about saying or reading the first word or letter or number the second or subsequent letter or number is perceived or spoken.

It is long-term process dominance.

You doubt that is the cause?

Not unusual.

After all, The International Dyslexia Association says: "Dyslexia is a specific learning disability that is neurological in origin. It is characterized by difficulties with accurate and / or fluent word recognition and by poor spelling and decoding abilities. These difficulties typically result from a deficit in the phonological component of language that is often unexpected in relation to other cognitive abilities and the provision of effective classroom instruction. Secondary consequences may include problems in reading comprehension and reduced reading experience that can impede growth of vocabulary and background knowledge." [1]

"The causes for dyslexia are neurobiological and genetic. Individuals inherit the genetic links for dyslexia. Chances are that one of the child's parents, grandparents, aunts, or uncles is dyslexic." [1]

Genetic predisposition indeed, but doomed to remain Dyslexic? Not in the least bit.

"Is there a cure for Dyslexia? No, Dyslexia is not a disease. There is no cure." [51]

Wrong, but right.

Dyslexia is not a disease but diseases are not the only things that can be cured.

When you get mad or react to a scare what is the first thing you do before you are aware that you did it?

Do you scream?

Do you jump?

Do you lash out?

Do you say things that later you wish you had not said?

All of us do.

That is long-term processing coming to a conclusion that needs output and sending the output to motion without a corresponding control signal from the short-term 'mind'.

What are the 'common signs' of dyslexia?

The list provided by The International Dyslexia Association is derived from *Basic Facts about Dyslexia: What Every Layperson Ought to Know* - © Copyright 1993, 2nd ed. 1998. The International Dyslexia Association, Baltimore, MD., and *Learning Disabilities: Information, Strategies, Resources* - © Copyright 2000. Coordinated Campaign for Learning Disabilities, a collaboration of leading U.S. non-profit learning disabilities organizations.

The 'signs' say absolutely nothing about what is really

happening, just what happens because of it and they are broken into younger and older person categories.

"High School And College:

May read very slowly with many inaccuracies.
Continues to spell incorrectly, frequently spells the same word differently in a single piece of writing.
May avoid reading and writing tasks.
May have trouble summarizing and outlining.
May have trouble answering open-ended questions on tests.
May have difficulty learning a foreign language.
May have poor memory skills.
May work slowly.
May pay too little attention to details or focus too much on them.
May misread information.
May have an inadequate vocabulary.
May have an inadequate store of knowledge from previous reading.
May have difficulty with planning, organizing and managing time, materials and tasks. " [1]

"Adults:"

"May hide reading problems.
May spell poorly; relies on others to correct spelling.
Avoids writing; may not be able to write.
Often very competent in oral language.
Relies on memory; may have an excellent memory.
Often has good "people" skills.
Often is spatially talented; professions include, but are not limited, to engineers, architects, designers, artists and craftspeople, mathematicians, physicists, physicians (esp. surgeons and orthopedists), and dentists.
May be very good at "reading" people (intuitive).
In jobs is often working well below their intellectual capacity.
May have difficulty with planning, organization and management of time, materials and tasks.
Often entrepreneurs. " [1]

Each of these 'symptomatic' issues is a result.

Not one is a cause.

If anything detrimental has a 'cause' it has a 'cure'.

Each of the attributes given to 'adults' in the list above that is positive in nature is a result of long-term memory processing.

Advanced long-term processing in the aural sense is usually responsible for the condition and can be a good thing when applied properly and a bad thing when it causes the 'mind' (short-term process) to wonder how that ever got out, or said, or out of order, or wrong or somehow not being 'aware' of an output the body has made.

The results of that inner confusion are shown in both lists above.

At the same site Dyslexia is "… a language-based learning disability…", "…The exact causes of dyslexia are still not completely clear, but anatomical and brain imagery studies show differences in the way the brain of a dyslexic person develops and functions…" [2]

NO, IT IS NOT A LANGUAGE-BASED ANYTHING.

It 'affects' language.

The differences in the way the brain of a Dyslexic person develops and functions are due to the overuse and dependence upon long-term processing and the lack of use of short-term processing.
It is a timing mismatch between long and short-term processes with a weaker short-term 'mind' output resulting in the ability of long-term to 'get through' unchecked.

There **IS** a Cure:

Take control of your brain.

By default, your 'human' brain has TWO outputs to motion, TWO evaluative processes.

Those TWO processes join at output to become ONE motion.

If one or the other is over-intense, or under-utilized the result is going to tilt to the stronger output.

Take the short-term awareness exercise in chapter two again and as many times as is required to gain control over your short-term process and do the following each and every time you read, write or speak from written or printed words or problems:

1: Remember YOU are reading or writing. YOU are creating or reciting. YOU are in control.

2: Find the 'mind' by concentrating on the task and not doing ANYTHING AT ALL without a CONSCIOUS decision to do it. If you have to walk your house without making a single step unless you have decided to make that step.

3: Read words by syllables. Read numbers with spaces.

4: If you wear glasses make it a habit not to make ANY decisions unless you can SEE YOUR GLASSES.

5: ANY TIME a 'feeling' or 'urge' comes over you to do or say something: STOP IT! THAT IS YOUR LONG TERM PROCESS TRYING TO DOMINATE YOU. That means: long-term memory is your record of the past and your past is making your decisions for you. MAKE YOUR OWN DECISIONS IN SHORT-TERM PROCESSING BY BEING AWARE THAT YOU ARE MAKING A DECISION AND THAT DECISION IS.

Chapter Five
Alzheimer's & Dissociation

There is a 'doctor' named Allison, who has come up with a typical subjective deduction about dissociation and calls the thing responsible the "Self-Helper".

That's fine to keep his or her 'patients' dependent upon someone who can help them reach the "Self-Helper" to exorcize their 'demons' but has absolutely no relevance to reality and in reality serves only to separate the self from the person being helped.

There are many other subjective observational deductions applied to problems faced by many and those 'observational illusions' are self proofing.

It is true that one can concoct a 'system' and make that 'system' fit just about any circumstance but it is another 'thing' altogether different to have a 'system' that is as finite and specific and (Occam's Razor) fundamental as the one employed in the topics offered at in this book.

Whether the 'condition' forms later in life or forms at a young age it forms at all as it is ignored into existence.

What is being ignored is the self.

Words like consciousness and self-awareness have been bandied around for centuries with no one person establishing a meaning to them.

Every person establishes meaning to terms used to describe something perceived by every normal human from that one human's perspective.

Such meaning is necessary to be established before any addressing can be accomplished to 'conditions' that are a result of both or either one missing.

Both dissociation and Alzheimer's revolve around the same 'system' occurrence but from different causes and form different paths of progress.

I'll quote a letter my wife has written to our daughter-in-law who is fighting an addictive condition of our her spouse. I believe this sums up the 'self' better than I can put it:

"Christi Critter's (she loves that nickname) life as a very young little girl and your life are similar in that you both have SO much potential. You are just not aware of it."

"About seven years ago there was a cute, chubby little blond haired, brown eyed precious 5-year-old girl who experienced life only by reacting to things. She even had an imaginary friend she played with constantly. She once 'peed' the bed and swore it was her brother even though he slept in his own room."

"She had not become self-aware.

Identifying the 'self' she had 'sensed' as being 'another being' who became the imaginary friend, she could not refer to herself as "I" as she did not know SHE was that separate being apart from the world that she perceived. She saw herself as part of the world and that 'being' she sensed as a 'different' part of that world."

"She was only aware of other things and other people until one day her Grandpa had her to extend her arm and then asked her 'Christi, (pointing to her arm) what is that?' She answered by saying, 'That's my arm, Grandpa.'"

"Then he pointed to her head and asked 'Christi Critter , what is that?'"

"She replied: 'that's my head, Grandpa'. Then Grandpa asked her… 'Christi Critter, who is 'my'?'"

"At first, she didn't 'get it'. But she quickly did and it changed a little girl who used to:

Laugh when someone else got hurt

Lie with every breath (well, not each breath, of course)

Had no remorse if an animal or different person was injured,

Saying she would "cut you" with a knife (she spoke to me in that one)."

"The world she was living in was nothing more than a world of reactions."

"Christi lived for about five years of her life exactly the way you have lived your entire life. I 'feel' your hurt. I really do!"

"After some deep thought (which was exciting to watch happen) she realized to 'whom' Grandpa was referring when he asked 'Christi Critter, who is 'my'?' … she looked straight into her Grandpa's eyes and smiled and said 'Me!'"

"At that moment our sweet little Christi gained self-awareness: the first step to a wonderful life."

"She has become caring, considerate, so loving, trustable, thoughtful and believable and now excels in school and especially with her classmates. She has interests in science and things that are larger than her and strives to understand them."

"Christi, now at the tender age of twelve years TODAY (as I write this) has what changed a little girl with no remorse into who she is today. She has knowledge of herself. She knows SHE is the only one like her and I'm very proud of what she will accomplish in her life."

Terms like 'consciousness' and 'self-awareness' are tossed about as if they were causes and not at all as if they are results.

The difference between the two can probably best be explained by a value.

If you win a card game by 1 point you could be said to be 'conscious' of the game.

It is the non-zero value of winning. Losing the game does not make you 'conscious' of the game.

It makes you 'unconscious' of the game both of which establish a 'condition'.

If you win a card game by 200 points you may be said to be 'aware', which is nothing more than a greater value of the minimum value necessary to qualify as the non-zero value of greater than the lack of any value.

Confused yet?

To be 'conscious' is to be aware of existence.

To be 'self-aware' is the increase in the same thing whereby one knows of the individuality of one's part of existence.

When a person does not gain the 'non-zero' conscious state, (science loves this term nowadays) it can be said they are a 'zombie'.

The same condition exists in the greater value of 'self-awareness' except we have real names for those conditions.

A more understandable example is a pond.

Sitting still it is a hole with water in it. Drop a pebble into the water and it becomes an undulating wave. Any wave at all stirs up the water to allow oxygen to enter it. No stirring (stagnant) and it becomes a cesspool where only bad things will live.

As with every observable condition of a human the first to

observe is the first to identify and normally the last to admit first impressions are never right.

Demons got the blame for what a person would say or do when no one understood the brain did anything.

That misunderstanding went so far as to condemn people who admitted they 'heard a voice' tell them something. Those people are still condemned.

That 'voice' was the long-term memory aural pathway coming to a conclusion for output based on what the short-term process had fed to it. It is the same thing as the 'conscience', happens AFTER a short-term evaluation or conclusion and not at all "con-science".

Both long and short-term processing are competing for output control. It is where the balance of the brain's system is observed.

The memory level most in charge is going to win but there are value collections, whole memories that when blended will establish an output greater than the control.

That is what is being heard inside the head. If the output was stronger still it would erupt through the mouth as most people will privately admit to having uttered things they were later sorry for saying.

When ignorance was replaced (in thinkers, not followers) with a more 'intellectual' reason a cause for that reason was not too far behind.

Since, at the time there was very little knowledge of the brain while the knowledge of the system and process employed by the brain has not been known until now, a 'condition' known to have been a brain issue needed something to blame.

It is the exact same process the initial observers needed to fulfill in naming the conditions 'demons'. It is also the same process employed by scientists who study the brain.

The brain is responsible for ah, let's see here... a whole lot of things. And since it is and since we can look with the help of an fMRI at the brain while it is resulting in those things we can see areas of the brain that do things too at the same time. We give those areas the distinction of being the 'blame'.

That leads to seeking the areas responsible for smaller things.

I just cannot wait until some brilliantly Ignorant Scientist proclaims to have found the 'miser' center of the brain. Some already have issued press releases claiming to have found the 'miser' gene.

IS's are now studying fMRI and CAT scan results to find the timekeeping center, the language center, the surprise center, the joke and laughter center, the emotional center and every other conceivable 'center' that must be responsible for the things the brain does.

It has gone so far in this categorization and compartmentalization that a researcher has now published a theory that cognition IS compartmentalization.

Ignorance spawns ignorants.

Dissociation Identity Disorder means the subject comprehends more than a single inner 'voice' or 'persona'.

It has been linked to potential 'childhood trama' (something has to get the blame) and that 'blame' would be partially correct but NOT the cause of the disorder.

It is a developmental factor. Two different people faced with the exact same childhood trauma will not have the exact same result.

The cause of the disorder is a short-term amplitude that is less or does not exceed the non-zero consciousness level.

It is the same condition a child perceives the world as before their short-term memory 'fills up' to reach the conscious balance of signal and amplitude that gives rise to the emerging 'self'.

It means, that while short-term processing is responsible for long-term memory in humans when that short-term processing does not control or 'affect' the memories being placed into the ordered long-term process it is nothing more than an awareness of long-term processes. NOT an awareness of self.

The 'self' IS that short-term 'looped' 'circular' memory process that is running far faster than long-term and input and is the collection of amplitudes we perceive as 'now' and as 'I'. 'I' in a healthy brain is not a condition of then. It is a condition of 'now'.

Long-term memory is our past.

Short-term memory is our 'now'.

When 'now' is always the same, long -term becomes 'now' and multiple perceptions develop that are no longer placed in logical order of time but are processed and continuously supported by 'now'.

When a child is growing up that child may have an imaginary friend as he or she grapples with the emerging 'self' that does not relate to past long-term memory.

Christi Critter had her's (yes, he was 'Chucky').

Our son Nicholas had 'flippy kid' which was in my estimation a delightful self diagnosis of what was then, a happy and glowing child.

When I was young I had me.

I would talk to me all the time for hours on end and wondered why I always answered and always knew the 'me' I was talking to and with was still just me.

I may indeed be 'older' now but the voice is still the same one that internally talked myself to sleep at 6 years of age.

When older that behavior would be called deep thought.

When younger it is considered 'weird'. Had this been a hundred years ago I would be locked up for admitting it.

Today I will just be ridiculed by those who do not understand it.

Those who do not understand 'it' are usually the 'visual' thinkers.

Short-term visually dominant people have a difficult time accepting that another person may not make judgments based on images.

Where dissociation emerges from the ignoring of nurturing the 'self' that 'nurturing' can take the form of abuse which tends to connect the emerging 'self' to the person doing the abuse, more than the short-term memory processing the abuse.

In D.I.D. the 'self' has not had the chance to be a contributor to long-term memory.

The conscious knowing of existence means the subject has a 'normal' output.

The subject is normally not aware of the other outputs as those other outputs are direct long-term reactionary signals permitted by a 'wave' of amplitudes that passes through the short-term process without being 'regulated' or 'changed' with the sense of the reality of 'now'.

Breaking those signals requires enforcing 'self' WITHOUT

referring to the other perceptions.

Referring to the other perceptions reinforces them and in many cases may trigger them.

Triggering them does not get rid of them. It simply gives them more to connect to and may cause the subject to 'become' another 'person'.

D.I.D. is unregulated long-term memory similarities caused by a non-active short-term memory process either in 'waves', built up over a long period of time in long-term memory supported as the same 'wave' with each 'resurrection' to short-term processing or in 'static' condition where short-term processing is not affecting long-term memory at all.

Alzheimer's disease is nearly the same thing except Alzheimer's disease is backwards and caused by a physical malfunction instead of a process malfunction and leads to a process malfunction.

Alzheimer's disease affects the connections of neurons in the return pathways of memory and like all other physical malfunctions it feeds on the weakest links first.

The weakest links in memory are the return pathway amplitudes farthest away from the source of input.

When it attacks long-term memory the subject will 'forget' more often until the subject cannot 'forget' what has not been remembered.

When it attacks short-term memory the subject will 'forget' the 'loop' responsible for being conscious and once disconnected (short-term is nothing more than a shorter memory in duration even though it is running far faster than long-term memory and takes up far more space in the skull) means the 'loop' no longer exists.

The more short-term processing accepts older 'conscious' memory the more it turns it into 'un' conscious memory (not the same as being knocked out) by not supporting at the same higher level of amplitude it was previously made with.

Memory itself is an interesting topic:

Today, the term 'memory' invokes digital computers.

Not until the computer 'had' a memory did the word

109

'memory' receive a use of its own.

Past forms of 'memory' were the printed word in books, the printed word on newspapers and the Kodak moment.

Memory in the brain is not at all like anything a brain has created. (Which would be a good yardstick by which to judge the next 'is responsible for' press release from fMRI and double-helix researcher reports. A building is not responsible for what goes on inside of it unless it takes control through defect.)

Brain memory is not actually the kind of memory we think of today.

Brain memory is not hanging around in some location in the brain.

It is constantly moving further and further away from the source of input.

That movement causes each pulse to reduce the amplitude of the memory wavelet. It sets up our concept of a passing time.

Memories that are not supported (even if the support comes from the same color or smell or taste or weight) are not lost they are smaller.

With each input comes another cycle of pushing older memories still older while supporting the most recently supported memories even more.

In long-term memory a connection to part of one memory can bring up a whole different topic.

In short-term memory a connection once reached will remain a connection unless it is ignored away.

When the Alzheimer's subject passes the previously 'conscious' supported long-term memory and starts to send only recently supported 'un' conscious memories to short-term processing, short-term reverts back to the non-zero value of consciousness until it drops to zero.

When that happens the rest of the subject's life is made up of their previous life with no relation to 'now'. Input stimuli that would normally connect to 'now' connect only to similar in the past.

Friends and family members become past friends or family members or if born after the time in which the subject is living are not known at all.

Alzheimer's is fatal because the long-term memory has faded into its past.

As unsupported long-term memories are ignored in short-term there is nothing to hold the long-term current and it digs deeper and deeper into the depth of the memory until it reaches the extremely small amplitudes of the body starting up and once 'remembered' can no longer run as no 'memory' of a 'heart' will not permit a heart to pump.

Everything is regulated by the brain and everything in the brain is regulated by its two main parts: Long and Short-Term processes, each with two main parts, visual and aural senses.

A healthy, well groomed human child can gain the awareness of 'self' early while an otherwise healthy, well groomed child can fail to develop the 'self' and in doing so can build up internal non-relational amplitude comparisons that emerge later in life as separate personalities.

Daydreaming is when the long-term process takes over due to the short-term process becoming equal.

"Getting lost" in a movie or a book is when the long-term process takes over due to the short term process repeating the same input.

In a book, it is the paper and type, in the movie or television show it is everything the eye is not focused on.

D.I.D. is not escape from a situation or a defensive mechanism. It is a condition brought about by the way the brain works and is one extreme of a single process:

'Self'.

Chapter Six
Comedy!
The Brains Of Williams,
Mochrie and Hicks.

Defining, Understanding and Possibly Ruining Comedy Forever

Disclaimer:

Before we start evaluating 'comedy' remember to read, remember and follow the directions that come with your comedy tools, knowing how to use your tools properly will stop needless injury and remember, there is no more important tool than these, groucho glasses.

As with all observation-based deductions there is a margin of error that only experience can provide a base line for.

There is also this little problem with being aware: "One's mind, once stretched by a new idea, never regains its original dimensions." -*Oliver Wendell Holmes*

Introduction:

The power went out a few days ago while working diligently on a project that rather required it and the thought occurred that the power 'going out' was a normal occurrence in this tiny part of South Carolina.

SCE&G has a wonderful telephone reporting system that always says the same thing.

I should know.

I've called it enough.

If you were to press 'redial' on my home phone the chances are you will connect to a computer at SCE&G.

It takes about an hour to get their attention, about another hour for their experienced work crews to realize there is only ONE location above ground where a power outage could take place, another hour to realize they did not bring the right tools to the site, again and what seems like less than 5 minutes for the application of the sophisticated engineering protocol.

If it was anything other than duck tape the power might stay on.

I raise that topic only because it was the culprit for this piece's postponement.

But now that I've managed to return to the original topic I wonder what I was supposed to remember.

When faced with that problem most of use will concentrate on the 'thing' we were supposed to have remembered and that it simply will not raise its head and we become frustrated and swear we'll be stuck thinking about it for the rest of the day.

Of course, we never think about it for the rest of the day, at least for as much of it as we are aware of it. We know that searching our memory for a specific missing part is like hunting for a haystack while measuring for needles. We'll probably walk right through it not noticing unless we were punctured.

Perhaps while mowing the lawn or cleaning up the bed or looking through a file cabinet or about to slip into sleep the answer will just sort of 'pop in there' and it will 'dawn' on us that we were aware of a topic without having immediately provoked the awareness.

Some topics take longer than others to formulate. It is those more 'important' topics that cause us to feel that moment of 'brilliant flash of insight'. And if the 'important' topic is removed in days or longer from the 'important' topic's first concentration we tend to 'forget' the original context and that 'brilliant flash of insight' takes on a mystical motif.

Add to the theme the textures of wonder that we, of all people could just 'sort of' find an answer to a problem that was either too long ago to make a difference or reasonably within time frames to be useful or even far fetched enough into the 'future' that the prudent observer would immediately assume a required protective confinement.

When we concentrate on a problem seeking a solution from memory we tend to think about a potential answer. A potential answer is not in 'memory' unless it is returned quite promptly after asking for it.

Since memory is a fluid process and each old memory, supported by new memory becomes a combination of both memories, connecting in some small way to other non-related memories it can take quite a while for the process once started by seeking a solution to find an old non-supported memory to merge with and just as long to bring it to the surface and will result in the 'insight' only if it connects to a 'now' event.

Each time we 'think' or 'concentrate' on a 'problem' we do so with our dominant sense. Visual 'minds' 'concentrate' visually. Aural 'minds' 'concentrate' aurally.

As a natural process the weaker sense still goes about its attempt to 'catch up' and is processing the same input as the dominant sense without our being 'aware' of it.

It is true with most of us that both types of thinking are evident in our short-term 'mind' but one of those two will have become dominant through either environmental manipulation (growing up in the early years from 1 to 8) or through mechanical mutation based in the genetic mixture of our parents or a combination of both, which is usually the case.

A way of 'seeing' this combination at work is by giving 'weight' or 'length' to each sense in both long and short-term memory processes and displaying the 'symbols' like this:

We represent the memory process with a vertical line: It is as 'high' on a scale of 0 through 10 as it is relative to its weight with its similar but opposing long-term memory process.

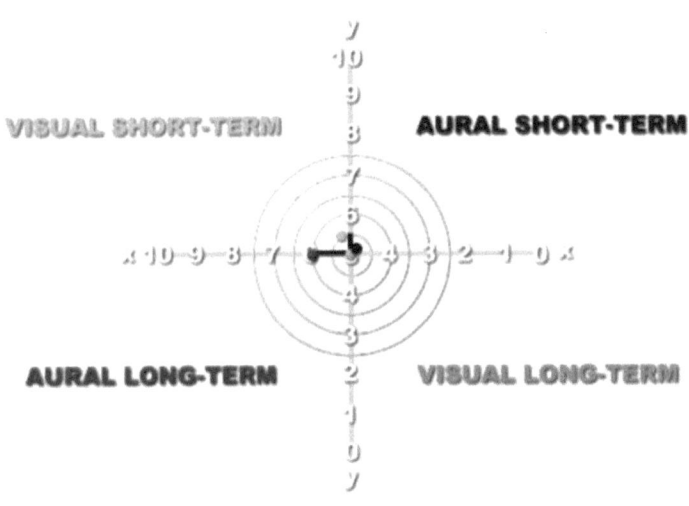

The mid level of '5' represents 'normal' and balanced mental processes. Anything above '5' represents an advanced 'intellect' or 'process' while anything below '5' represents a decreased 'intellect' or 'process'.

Our subjects are all male so we'll assign the value of long-term to the single 'whole' value above normal in the long-term aural sense.

Within the range of ½ of normal the brain is still capable of affording its body a relatively normal life as long as all senses are balanced, which does not mean equal: (balance is between long and short-term process, not the method by which the amplitude is regulated) a high short-term can be balanced with a reduced long-term and a perfectly normal person will emerge that contemplates the sense dominant while having difficulty accessing the sense subservient.

The left graphic depicts a 'normal', 'conscious' male human being. It shows a whole of 1 (amplitude equal to the system amplitude in range) increase in long-term aural process (which is as close to 'awareness' the long-term can get as it is not able to set up a complete 'loop' of feedback return due to the very low amplitude of the oldest 'memory' wavelets) with a non-zero value with amplitude less than the system value in range in 'conscious' level of short-term aural while visual resides at a 'normal' long-term and aural short term is shown as normal, less than the parent long-term value amplitude.

·1/2 We wind up with a symbol for a normal thinking male human that shows a near balanced system.

·1

Long-term is represented by a left horizontal line indicating aural long-term 'presence' while short-term is represented by a vertical line indicating aural short-term presence.

Such a 'balanced' condition has ramifications:

Advanced aural long-term process results in recall of concepts that relate to stimuli that relate to stimuli, which become connected to the concept because the concept connected to them.

It lacks equality in visual processing in both short and long-term memories and as a result will not have much interest in appearance, orderly arrangements or the finer 'arts'.

116

Will lack the visual gracefulness of motion and replace it with aurally dominated brute   force to accomplish a task that could have been 'clumsily' accomplished.

And it will seek to increase its visual deficiency through visual stimulation which could, if let fester, result in a 'need' for visual stimulation which could lead to a 'need' for a normally visually controlled process of motion to 'need' to over-ride the resistance to action through excessive and oft' times abhorrent actions and re-actions.

The 'elite' would view the person today as a barbarian, far too many wives can relate to the character while the 'trait' of aggressive and dominant and 'alpha type' qualities associated with our ancestors begins to make sense.

Without that brain created advantage, the Homo Sapiens type species collection most likely would have followed each predecessor into oblivion.

The stage of the dominant personality's balance of processing sets the stage for the form of culture, the form of government tolerated and the form of curiosity growing within that culture.

With each new concept of information collected by this 'alpha' male dominated culture comes a new question for the parts of the concepts are not complete.

Once that process starts it grows and grows until the desire to fulfill the concept, met with the unknown can be compromised by acceptance of 'theory' as 'de facto' in every single defense of a 'theory'.

A 'theory' that is not 'theory' cannot be defended. It is a singularity and has no relation to the parts that it describes other than collecting them in its specific method or order in order to explain them.

When the dominant male is not the 'alpha' but rather the 'beta' the culture represents the opposing symbol.

Such a 'balanced' condition has its ramifications as well:

Advanced visual long-term process results in recall of images that

relate to stimuli that   relate   to   stimuli,   which   become
connected to the image because the image connected to   them.

It lacks equality in aural processing in both short and long-term
memories and as a result   will have much interest in appearance,
orderly arrangements and the finer 'arts'

Will lack the aural precision of logical motion and replace it with
visually dominated grace and fluidity to accomplish a task that
could have been illogically accomplished.

And it will seek to increase its aural deficiency through aural
stimulation which could, if   let fester, result in a 'need' for aural
stimulation which could lead to a 'need' for a nor-   mally   visually
controlled process of motion to 'need' to over-ride the resistance to
action through excessive and oft' times abhorrent actions and reactions.

A completely balanced, not dominated society would look like
this:

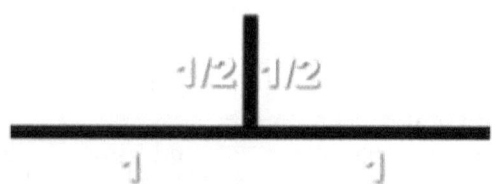

And result in chaos, lack of respect for the rule of conceptual law,
always attempting to remove parts of images that do not seem to fit
and the abundance of visual stimulation for gratification, provided by
aural stimulation to create, and aural stimulation for order, provided
by visual stimulation to reproduce, accomplished by a repetitive 'beat'
nearly equal to the first level of processing in the brain and result in a
near 120 beats per minute rhythm that is both comforting and
fulfilling to both types of thinker.

It is the condition our original ancestors put up with, as their
knowledge level was nearly zero and every thing that had order was
special while everything that had order noticed only over a long
period of time was worthy of deeper curiosity and every deeper
curiosity brought inability to comprehend, which festered into the

'need' to find a fit to the picture or the missing part of the complete concept.

It is why all early Homo sapiens looked both up and out. Up, was vast and larger therefore more important.

Out, was also up, therefore started important, but was deeper than up, therefore gained a higher degree of importance.

The larger things 'up and out' were viewed as special and always there. But that is historical and I digress.

The Brains of Williams, Mochrie & Hicks:

Robin Williams was born as the son of an executive of the Ford Motor Company on July 21, 1952. His early childhood is familiar as his parents were older; the 'children' were grown. The solitude of a forced 'mind' is able to seek its fulfillment in things that are conceptual and do not need mass by which to take hold.

Some output that process in the 'talent' afforded to them to capitalize.

Mine, was music as my older brother (much older brother) was excelling playing the Clarinet in High School band and orchestra and even though his exploits amounted to great admiration for his accomplishments, it is a hard goal to have forced upon one's formulative years, especially when it has nothing to do with the Clarinet one must take lessons on, or the subject of macro college acceptance, one must aspire to, and never quite ever make the grade.

Mr. Williams suffered a different form of deterrence. The constant moving around made him the 'new kid' in every snotty filled 'private' school he was placed in. His grounding to conceptual logic found the visual stimulation of Jonathan Winters' albums fulfilling as Winters was (still is) a master at manipulating concepts from visual to aural and back again.

The somewhat "pudgy child" [1] was the target of ridicule in schools he barely knew.

Acting upon his formulated conceptual abilities formed

from within a closed space of his own 'mind' Mr. Williams began to use that ability to rid himself of ridicule.

A normally calm and caring individual deflected pain with the weapon he used best: his greatly advanced long term visual processing.

Advanced comes in three methods:

1   By increase in clock speed,

2   By increase in feedback return clarity

and:

3   By an increase in amplitudes.

In relation to long-term Mr. Williams made a connection to visual short-term as he used and practiced using comparative relationship alternatives.

If the concept is a 'cow' the aural comparative relationship alternative is a four footed animal that can relate to dinner last night, which can relate to the argument over whatever that was, which can relate to the previous argument, which still festers and the person can come across as a tad irritated thinking about a 'cow'.

If the concept is a 'cow' the visual comparative relationship alternative is a two legged animal that can relate to the self-image the perceiver may have of their own body, which can relate to dinner last night and that look he gave at the second helping of potatoes, and the last argument will fester as the person exhibits a tad irritation about the 'culprit' who brought that over-weight condition to her attention due to his always being a tad irritated.

Of course, other alternatives are available if one cares to seek them out.

It is the process of taking what is and applying what has a comparative relationship. In other words, it is 'dark' comedy.

But only to the observer unless one is using it to deflect a bad situation and convert it into a good situation and should that person be the perpetrator of the 'dark' comedy it will most often come as just a surprise to the deliverer as the receiver. It is a wonder to watch Robin Williams unleashed.

It is life viewed with fictional connections.

'Light' comedy is the opposite. Fictional connections viewed with life's perspective.

The processes are exactly the same as their collective opposites.

The collective opposite of 'dark' comedy is 'light' drama.

The collective opposite of 'light' comedy is 'dark' drama.

They all result in the same causal process exhibited in humans as either laughter or fear.

Both laughter and fear are the collective opposites (or co-relational opponents) of the opposites of output at all: 'shock' and its degrees all the way down to chuckle in the face of the slightest fear or bruised sensibilities at the use of a term or a visual implantation.

It is one of the four types of human 'tilt'.

When visual processing in short-term is relied upon to connect to a known working solution from an aurally dominated long-term it begins to creep up to a near equal level of awareness in amplitude and the comedy presented is not only quick and funny it is physical in a duplicative way as Robin Williams is the reigning master in residence in our society.

That condition would tend to lead the person having it to identifying other weak or seemingly 'oppressed' conditions or situations and would rely upon the same logic to address them.

I suffered a result of that condition not too long ago.

In the process of evaluating the brain's physics and in the process of communicating that evaluation to a 'mixed' audience the process of employing both types of thinking in order to reach a general understanding required me to go through the same process but in a far shorter period of time.

It resulted in my becoming aware of the 'awareness' in short-term visual processing rather abruptly and as all first events to a brain are, there was literally nothing 'in there' to connect it to so it 'tilted' in fear and my long-term process took control of my mouth which resulted in a rather embarrassing admission of not being aware of that level of control until the resulting barrage of well deserved corrective verbs were returned.

121

That shall not happen again.

Others suffering it much later on in life after its becoming 'aware' slowly over a lifetime seek 'fixes' to personal problems whether those personal problems are personally caused or not. It works the same way whether the person is aurally or visually dominant.

Those fixes can take the form of drugs, alcohol and meditation as well as whatever else a person can connect to a deflection of reality.

Robin Williams' Brain Before Realization:

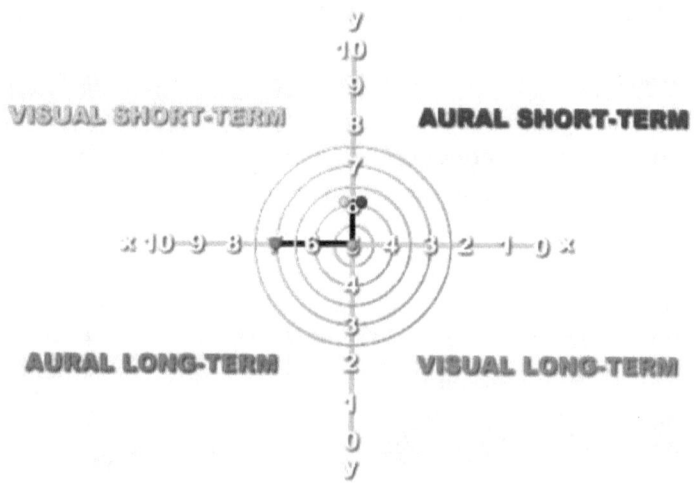

1-1/2+1-1/2+1 = 1 a human 'conscious' brain.

Still unbalanced, this aurally 'advanced' long-term and mutually advanced short-term brain caused its owner (or resulted in its owner) to attempt experimental alternative co-relational opponents.

When the degree of disgust at the lack of results exceeds the degree of fulfillment received by the normalcy of the condition a 'wake up' occurs in some people. Robin Williams got his 'wake up' about the time his friend John Belushi died.

It resulted in a deeper short-term evaluation based in the

primary sense (aural) for conceptual purpose and the use of the short-term returned aural to control and nearly balanced it to long-term erasing the 'normal' of being depressed with the original 'normal' of being aware.

Of course there are those undoubtedly regrettable recurrences of emotional responses and the one remaining visually based viewpoint of politics Mr. Williams is more apparently disgusted with than agreeable to: and the chart changed.

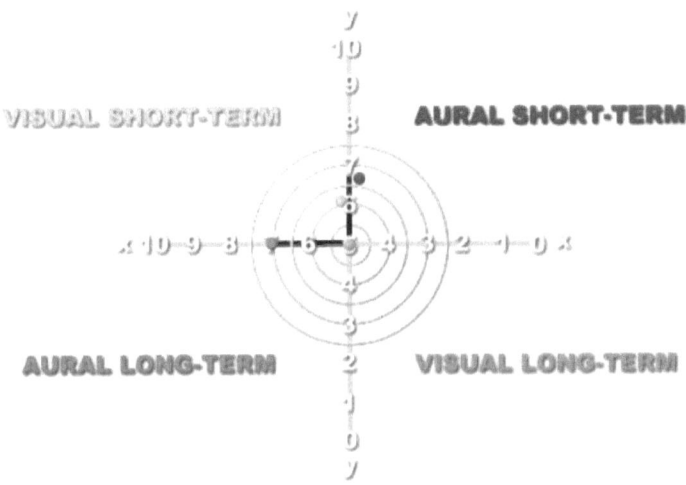

Applying the 'equation' to this collection of attributes:

$$.5-2/2+2-1.75/2+1.75= 1.5$$

A "1.5" human self-aware brain:

It has made him return to his first goal of serious acting and has resulted in some amazingly insightful portrayals and some confused career selection replacements for alternative co-relational opponents.

Robin Williams is indeed a beautiful mind as the 'mind' of Robin Williams is Robin Williams and I would not be surprised if he did not already know that.

Only a direct interpretation of this observation evaluation can authenticate or amend its deductions from such observations but I greatly doubt Mr. Williams, even with his intent interest and curiosity in 'artificial' intelligence would ever find the book let alone find time to read it.

Colin Mochrie hails from Scotland when on November 30 1957 Mrs. Mochrie expelled Colin as his welcome into the world.

His family moved to Montreal in 1964 then on to Vancouver in 1969 and young Colin continued to be pulled along with his family as his father's job as an airline maintenance executive set up the same 'new kid' situation Mr. Williams went through in frequent moves.

His young 'condition' is described as: "He considered himself a loner during his childhood days due to his shyness, a trait that he claims he still evidently has, and the frequent moving.

At some point, as a child, he had dreamed of becoming a marine biologist or a chef. However, at 16, he was dared by a friend to join a High School play where he had the part of an undertaker. It was then when he got his first laugh (by splitting his pants) and afterwards, he "craved for nothing but more laughter."" [2]

In his 'auto-biography' at the same resource Mr. Mochrie does not refer to himself as a person he refers to himself as a passenger: "I went to theater school for 4 years, then luckily managed to get work. I got involved in Improv through the Vancouver Theatresports league. I moved to Toronto after Expo 86 and got involved with The Second City. A few years later I moved to L.A., hated it, and moved back to Toronto. I married in 1989 (to Debra McGrath) and have a son (Luke). I was with Second City for 3 years (a famous North American comedy theatre). I just finished filming the part of Grady in the TV show Once a Thief. The show was cancelled two days after. I'm sure its just a coincidence. I had a small part in a movie called The Real Blonde with Kathleen Turner and Matthew Modine. Of course my most famous role was in the 3D Space Epic Space Hunter: Adventures in the Forbidden Zone. I capture Molly Ringwald and utter the immortal line "you can ask the Chemist"." [2]

Mr. Mochrie is so funny and so fast and so logical as his long-term aural process is dominant over the entire system while his short-term visual 'mind' is slightly over non-zero and his aural short-term 'mind' is normal.

It is a very similar condition to Mr. Williams but without the added incentive of defensive use of inner thought Mr. Mochrie is aware that he is 'in there' but not aware of what is 'in there' is him.

If he ever does become aware of himself he will not be quite as quick or quite as creative nor quite as unpredictable. It is a situation that Mr. Williams has lived in since he became aware of 'self'.

Not having the complete awareness of self is a job requirement for comics. Wayne Brady, starring with Colin Mochrie, Ryan Stiles and Drew Carey on the American version of Who's Line Is It Anyway? very rarely allows his short-term awareness to overtake his long-term abilities and is able to 'rap at will', so to speak in either verbal or physical progressions. Mr. Brady is short-term aurally aware, visually long term advanced but advanced further still in long-term aural, (makes for an interesting talk show host): so advanced aurally in long-term that his short-term does not rule the output.

He becomes the part he portrays.

He portrays any 'thing' able to be imagined as a conceptual 'thing' and by collecting parts of images is able to recite those images in great detail with rhyme and song.

What he cannot do well is make a long-term 'funny' relationship from an external visual image.

Mr. Brady does not take part in the 'describe the video' segments of the show and if one watches his performance in skits where he responds to unexpected visual stimulus one will get the impression he finds that point to be where his powers of instant verbal recall are lacking.

(For everyone else, they would be normal!)

Mr. Mochrie applies 'light' comedy by injecting everyday things into seemingly impossible to match relationships and does it so quickly that shock of response alone would garner him merit but it is the depth of his long-term recall that gives him that response and it makes him feel alone yet not fully aware that we all are.

Mr. Mochrie's brain charts like this:

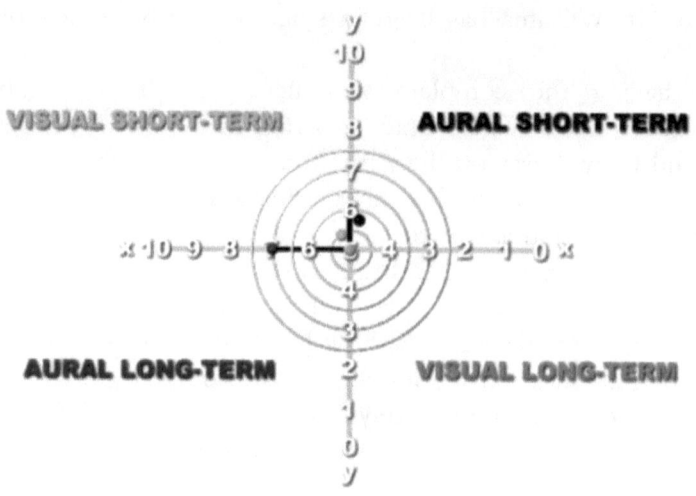

And is shown like this:

.5-2/2+2-.75/2+.75= 1

A "1" human 'conscious' brain.

The equation combines the two left halves with the right half to arrive at a 'standard' degree of 'awareness' reference.

Only a direct interpretation of this observation evaluation can authenticate or amend its deductions from such observations but I greatly doubt Mr. Mochrie would ever find the article let alone do I think he should.

A name Americans might not be familiar with but Europeans, especially the Brits will recognize is Bill Hicks.

Bill Hicks was perhaps one of the greatest truly 'dark' comics to ever have landed on stage.

Mr. Hick's parents, especially his mother (I can say that, I have spoken to her) are rooted in old time religion hailing from the Valdosta, Georgia area where on December 16 1961 little Bill burst forth to start to make sense of all of those muffled sounds experienced in those months of preparation for self support.

He too, moved more than 'normal' and at the age of 7 landed firmly in Houston, Texas where his family took up roots in what Mr. Hicks would refer to as a; "strict Southern Baptist ozone." [3]

Bill's aural evaluation short-term process made a different change in him than it did in Mr.'s Williams and Mochrie. It became dominant over long-term as he evaluated those values he could not value and the rules he could not find logic to support.

Having come to a sense of 'self' early on in his life and having protected it as well his philosophy is summed up by his biographer Paul Outhwaite, as "we are all one consciousness". [3]

Bill Hicks saw the 'world' as 'out there' and his 'self' as 'in here' and insignificant. His musings were meant to enlighten and correct misinterpretation of logic through observational subjectivism. His methods were based in what it took to get noticed all of his life: sensationalism.

His sensationalism was meant to invoke emotional response, so his comments would capitalize on them instead of having to blaze their own path to the audience's many different perspectives. He created his own relational subjects and filled them with alternative relational comparisons. The more 'themed' his subjects were the more definitive he could get as a "Jonathan Winters of aural perspective" or a near "Woody Allen" of the untainted.

He saw life as "just a ride".

"At 11.20 pm on Saturday 26th February 1994 he died in Little Rock, Arkansas, buried in the family plot in Leakesville, Mississippi.

At the memorial service Hicks' brother read out a piece Bill had written and requested be read:

"I left in love, in laughter, and in truth, and wherever truth, love and laughter abide, I am there in spirit." Bill's spirit then floated up into the cosmic one consciousness where he continues to enjoy the ride throughout eternity and infinity." [3]

Bill Hicks is a career and collection of work one would be most welcome for having found. His comedy is not reactive and is not for the faint of heart.

The Brain of Bill Hicks:

A "1.875" human self-aware brain:

Only a direct interpretation of this observation evaluation can authenticate or amend its deductions from such observations but I do not doubt Mr. Hicks would have found the article and torn it into its parts and derived its main concept and used it to poke fun at the stupid little things people do.

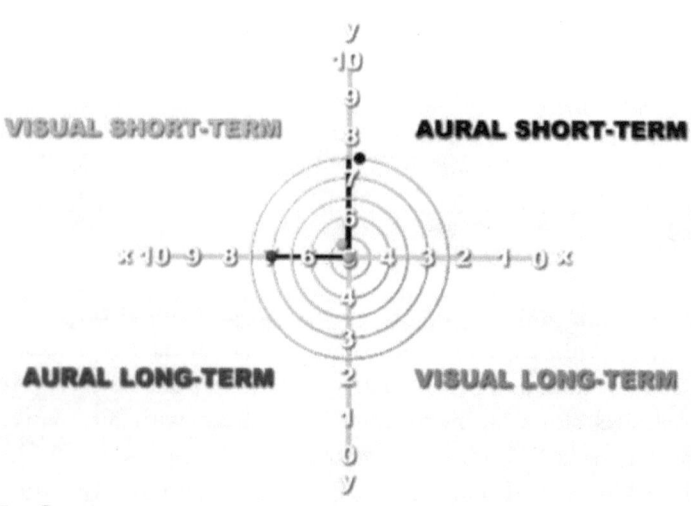

In Common:

Dispel any thoughts one might have in comparing degrees of self-awareness.

There is no 'stigma' with a degree of self-awareness.

It is NOT an intelligence test and it is NOT a method by which to discriminate things apart when they should be common to each other and kept together for each other's sake.

What these three giants of comedy have in common is the ability to portray a concept so other people, regardless of their mental dominance and viewpoint will understand it. That commonality is a process. Not a 'thing'.

Oh, that's right. I was going to ask the reader to pay close attention to the words and connect those words to the concepts they create and allow those concepts to brush by already known concepts instead of bounce off already known beliefs and later to allow the new concepts to emerge just once, in an evaluation of a condition or situation but one has already read the chapter.

Chapter Seven
Understanding Sexual Preference

Before the end of this chapter, I wish to apologize.

I feel bad, in that it has taken this long, for the ability to reach fruition, by which the publications during 2003 could be written.

During the Spring of 1993, after having spent seven months pouring over test results of a circuit design, and attempting to find any possible way in which the interpretation of the mathematics could be in error, I made a self-aware decision to not concentrate on the main topic of the technology.

Curiosity, the result of a repeated void response from long-term memory, can be a very stimulating experience, when one is curious about demonstrative applications of a singular system.

During an intense period of time in May and early June of 1993, I wrote many speculative papers on varied topics of potential for a model in which the mathematics could be demonstrated. The single criteria for acceptance of concentration was the universal applicability to a potential understanding.

The topic finally decided upon was the brain.

The brain, (human level), is a perfect architecture by which to start an understanding of a universal system.

Everyone has one.

Everyone who would potentially understand would be capable of understanding.

I spent 10 years (sporadically, as I was not able to devote anymore than a cursory amount of time to the project) cranking out absolutely true papers of different aspects of the brain.

Those papers, were not at all easily understandable and those were thrown away, yet the papers of the second reincarnation of the topic remain on this site, more as a reference for those willing to attempt to make sense out of an amateur's ranting, than anything else.

The self-aware decision has been a problem as well as a benefit.

It was impossible to step up to the plate in any scientific forum, whether it was as ridiculous as UseNet or as superlative as a mainstream science publication.

How in the world was a person to mention a 'different' energy charge to a science that had never seen one?

A few years back science finally saw one.

It got me back on track.

The time had come where a new discovery (to science) was in

its true understanding-infancy, and no peer-groups had formed strong enough around a single acceptable understanding of it, to cause much of a resistance, to understanding what was to be understood, instead of what could be understood.

The self-aware decision came back into play when in my enthusiasm, to dive back in to the physics I love to swim in ,as it was still not time.

All of the writing and explaining done prior to 2002 regarding the brain, was not understandable, and in most respects, was rather boring as my rather obnoxious aural short-term dominant process, was getting in the way of saying and drawing the right images at the same time.

It was not until I met Fred (not his real name), that the concept of saying things in the way the listener will hear them, turned on a light in an otherwise cobweb infested topic.

I spent a good deal of time on the phone with 'Fred', who is a short-term visually dominant, long-term dominate male, describing the concepts I was portraying, in a way he would be able to gain an image from, that matched the concept I was trying to explain.

At first, it was quite frustrating for both of us.

But after a short period of time 'Fred' began to view my way of thinking, and I began to view his way of thinking, from the other's perspective. Our conversations became less and less cumbersome, and my ability to converse, in a way reducible by visually dominated short-term thinkers, became not only better, but easier as well.

It became so 'easier' that I let it get the best of me, which is covered in a different paper and I do not have to go into those details again here. Whew!

So, I began to set about the task of making right what was not wrong, but was incorrectly worded all those years ago.

If I had known then, that all those years of frustration were not necessary, it is quite believable, that I would have opted for the easy way out, and jumped straight to the point, which science was not in a position to have taken seriously, let alone accept.

At that time.

Once again, like an iggit, I decided to cover the same topics covered in the first round of papers but do so in an understandable manner.

Of course, there are knowledge issues in the other papers and some issues that contradict accepted knowledge but those are for those papers.

When I realized that I had managed to duplicate the same topics I knew there was something most important I was missing.

For the past week (as I write this chapter), I have been laboring over the curiosity of what that missing issue was.

In reading over the other papers of 2003 at this site one will find references to many sub-topics inside of each main topic paper.

The sub topics do not deserve papers of their own.

I kept wondering, that of all of the topics covered, and all of the explanations given time was going to be the largest culprit and if I had just thought it all the way through 10 years ago then now, perhaps by now, a splattering of the knowledge would be slowly sinking into the treatment of mental disorders and psychiatrists would be doing their patients good instead of lip service and pills.

I do not feel bad that it will undoubtedly take far longer for science to drop its illusions and come to its senses. I have no responsibility for that.

Then, just a couple of days ago (as I write this piece), I received an email from a man looking for a specific programming product similar to one I had managed to crank out earlier.

In that email conversation, where I found him to be quite kind and intelligent he referenced the domain the software would be used at and I checked it out. It dawned on me then, that what was missing was what could do the most good.

But, I didn't have a reference point.

Then, I did.

MSNBC did the right thing, one thing too late.

Michael Savage is not a very smart man. I judge 'smart' on how much control over output the short-term process has over the long-term process and Savage's remarks, hastily excused as 'not intended to be on the air' proved his instant response was long-term based, and it continued, according to the news reports. Trent Lott knows that problem as well.

Not only did Savage resort to long-term based slurs in an attack he apparently did not have the short-term intellect to deal with, he kept it up, meaning it was not only long-term it was his anger venting with his perception of the truth.

From his own web site: "It was not meant to reflect my views of the terrible tragedy and suffering associated with AIDS."

[The incident that resulted in his firing began innocently enough. Savage was taking viewer phone calls about airline horror stories, and a male caller began talking about smoking in the bathroom.

(Experienced helpful hint: delay processing delays both voices, MSNBC cut off the caller's voice only.)

"Half an hour into the flight, I need to suggest that Don and Mike take your ..." the caller said, before he was cut off and his words became unintelligible.

"So you're one of those sodomists. Are you a sodomite?" Savage asked.

The caller replied: "Yes, I am."

"Oh, you're one of the sodomites," Savage said. "You should only get AIDS and die, you pig. How's that? Why don't you see if you can sue me, you pig. You got nothing better than to put me down, you piece of garbage. You have got nothing to do today, go eat a sausage and choke on it."

He asked for another phone caller who "didn't have a nice night in the bathhouse who's angry at me today."

These bums "mean nothing to me," he said.] MSNBC Referenced Above.

I tend to believe the source with the tape in stock.

But as much as I trust the tape, I distrust the typical radio stunt this whole situation really was.

It was one radio show against another one using MSNBC as its battleground just as it has done with Fox News, Larry King, CNN and many others.

Savage knew it immediately, but instead of addressing the topic, which was Don & Mike's radio flunky, Savage addressed his ignorance.

Also in the tape, Don And Mike identify who the caller is, that he is advertising their radio show and that they claimed MSNBC had changed their calling acceptance policies and procedures to stop that caller, yet he got through?

MSNBC may have set it up, knowing who the caller was by

his phone number but one would have to believe Don & Mike's claim MSNBC was ready for Bob Foster. The truth? I don't know.

What I do know is, it is a very old radio scam and the very old (sorry, I can say that) Savage did not fall for it.

He caught the caller the moment the Don and Mike names were said (it just happened to have been the moment after it was said).

His problem started when he unleashed his true feelings of ignorance on a person who was calling for a different reason, and he knew it.

And he kept it up.

Anger is based in something. In this case in listening to the tape it appeared Savage's ego got the best of him and his 'id' won out.

Try as I might I cannot find the reference to the audio I heard on the radio today about this topic.

Neil Boortz is the only speaking human being on the air in Charleston, South Carolina during the early mid-day and my car radio wound up where the signal was human.

During his program he discussed the Michael Savage issue, completely ignoring the radio stunt 'angle' (which is exactly what it was) and in response to a male caller referred to something he had heard Savage say earlier. (Which means this is hear-say of hear-say, but good enough of a reference point excuse for a paper as I've ever seen.)

Boortz claimed that Savage said (sounds like a grade school discussion) [recalled, paraphrased] '...until the time came that someone showed him that homosexuals were born that way he would consider them to be....' [lack of recall, unable to paraphrase] some form of expletive slur.

Hello Mr. Savage.

Let me introduce you to the brain and the way it thinks.

Normally, (which means the majority) of human brains are indicative of their gender.

Females, 'normally' are visual long-term creatures, with either aural or visual short-term processing. Males, 'normally' are aural long-term creatures with either aural or visual short-term processing.

The 'masculine' traits all stem from aural dominated long-term while the 'alpha' male dominant, head of the gang, born leader is the aural dominant short-term male with aural dominant long-term memory. The slower the short-term processing is in that instance the more 'macho' they are.

The 'feminine' traits all stem from the visual dominated long-term while the very feminine submissive, born victim is the visually dominant short-term female with visually dominant long-term memory. The slower the short-term processing is in that instance the more 'valley-girl' they are.

Just as it is possible for the aural and visual to vary in short-term of 'normal' gender 'specific' persons, so it is possible, and in fact the cause of the opposite mental 'preference'.

'Normal' means the most accurate fit to the architecture.

Females are visual long-term, as their evolutionary role of the 'gatherer' and 'child-bearer' require a visual interaction, where the conceptual aural long-term female, would never suffice.

If females were not visual, the species would not exist today.

Visual receives in order to create.

Males are aural long-term, as their evolutionary role of the 'hunter' and 'protector' require an aural interaction, where the visually controlled long-term male, would never suffice.

If males were not aural the species would not exist today.

Aural creates in order to see.

Every aspect of the brain is a balancing act.

Each teeter ,causes a totter (whatever that might be, it sounds good), for every action there is an equal and opposite reaction, and so forth, and so on.

Female is balanced with the male, where female is the default state of all potential humans.

Aural is balanced with the visual, where female is visual default and male is aural default.

There are variations in all cases.

Variations are not bad.

Variations are not evil.

Variations are normal BECAUSE variations balance OTHER variations.

If 'Fred' can be long-term aural and short-term visual then another male can be long-term visual and short-term aural.

Those other males will feel feminine but act masculine. (Visual short-term would be acting feminine).

The same applies in the other gender.

If 'Freda' can be long-term visual and short-term aural then another female can be long-term aural and short-term visual.

Those other females will feel masculine but act feminine. (Aural short-term would be acting masculine).

A normal short-term process compared to a normal long-term process, will depend for control over which process is first to reach maturity and become the 'self' the person perceives as 'I'.

If the person has not managed to find the 'trigger' or the importance of 'self' the amplitudes generated by short-term processing will be just above the 'non-zero' value past equal. The result will be 'consciousness' but not 'self-awareness'.

Those persons make up the majority of humans.

True 'self-awareness' has been the target and subject of mysticism for thousands of years when it is really just a normal human trait that we have neglected to teach the use thereof, because we have not understood the brain at all.

There is nothing bad about a person who finds the mental desire to feel 'normal' internally, with 'normal' being different for them than it is for you.

The sexual reproductive process of humans has been considered a bad topic as it causes embarrassment to those who view it with a special sense of purpose.

That is not the topic of sexual preference.

The topic of sexual preference is for the contentment and feeling of 'normal', that matches the image or concept the 'self' has arrived at, through repetition of input.

If the input being repeated is the long-term memory (self-awareness is very low), then the image or concept of 'self', will need to match the dominance of the processing type.

If that processing dominant type is contrary to the gender's default condition, the 'self', will not feel like the right ,'self'.

Admitting the short-term's perception of self is based in the long-term's perception of self, is a good starting point to finding out if the feeling of 'normal', expected from what memory says should be 'normal', is based in a condition of birth or a condition of confusion.

If the person is female and their long-term memory is male the 'self' they will perceive will feel fulfilled as a lesbian. They are born that way.

If the person is male and their long-term memory is female the 'self ' they will perceive will feel fulfilled as a gay. They are born that way.

Where Mr. Savage and most other sexual bigots miss their mark is in the simple act of sex.

Sex, is a reproductive act. It is, luckily, pleasurable or the species would not be here today.

Just about all brain containing creatures find 'pleasure' to be a good thing and seek it out.

The sex act has nothing to do with the preference of the sexual preference since sexual preference is mental gender preference.

The sex act is the only physical manner in which love or lust (either extreme of pleasure) are able to share in that emotion or that misjudgment.

Not only are 'true' gays born that way, Mr. Savage; so are talk show hosts.

I know, I've tried it. One single show at one single station in Phoenix, Arizona.

All set to take on the callers only to find out the previous four million weeks of that time slot were covered by local high school basketball (which no one ever listens to) and my impromptu 'fill in' was not only unannounced, it was a 'secret'.

I was prepared. I was ready to fill the entire show with me talking. just in case no one found me interesting, while I took the chance to examine the, then current events , in a manner of pure logic.

I had no calls.

I struggled through the first 20 minutes of the show, into the first commercial break, when it dawned on me that long-term thinkers make good talk show hosts.

Talk show hosts these days are not at all about their topic, or their 'cause'.

They are about sensationalism, getting the calls by exciting whatever emotion they can manage. The point matters, not the consequences or the victims of it.

When the commercial block was complete and my 'bumper' music started I began to play with the producer and stopped looking over my notes, and let the long-term take over with occasional disagreements from short-term.

I have the tapes.

The one call I did receive during that show was from the assistant producer pretending to be a caller from another room in the station.

I think he felt sorrow for me.

I didn't like that when I found out about it after the show, which is probably why I never fit in that industry.

Don and Mike treasure the deception.

It makes me sick.

So as simple as the explanation is of what causes the mental 'state' to exist of an internal feeling of needing the 'self' to not be a lie, and as much as religion preaches against lies it is amazing how many will read this paper, find out about what they think it says and still come to the conclusion that it is worth ignoring.

There is a cause of sexual preference, and it is active in every condition of the brain.

It is 'normal' for the gender if the long-term processing is normal for the gender. It is 'normal' for the person whether it matches the gender or not.

Radio talk show hosts need to be long-term thinkers, with slightly advanced short-term, and should be aural long-term and visual short -term.

They can paint a picture from a long held belief and listeners will 'see' that picture while listeners who cannot 'see' the 'picture' will relate to the 'concept'.

It is what makes the hardcore, shock-jock fan what he is.

If a radio show host is far advanced short-term and is visually short-term dominant, male or female, they will not only make great radio salesmen but they will advance into management, which is why radio as an entertainment medium these days has about as much quality, innovation and entertainment value as a sewer's sludge.

If a radio show host is far advanced short-term and is aurally short-term dominant, male or female, they will not only make great on air talents that require thinking to listen to (therefore they will be entertaining to the intelligent and qualified target listener) but they will make great program directors and horrible talk show hosts, unless they know how to let go of the self. They will make it into station management only if a natural disaster suddenly kills the general manager., and no one else is in the building.

There is no difference between what causes a radio talk show host's talent and what causes a gay male's internal feeling of not being who they know they are until they admit it.

Now, you know.

Find another target for your sensationalistic searching consultant to embrace.

As for my apology: It still applies.

But now, not with as much remorse.

It took a long time to reach the point where explaining it at all, was possible.

For that, and any part I played in hampering the event, I am sorry.

If only people would know and learn.

Perhaps it might not take another ten years to start easing the emotional pain of confused brains.

Knowledge is only knowledge, if it is known.

Chapter Eight
Facing Your Brain – Taking Control
Overcoming Depression, Withdrawal & Imbalance

Drugs work on synapses, which are chemical translators of wave amplitudes by either raising (amplifying) or lowering (resisting) the translated transmission.

"Neuropharmacology is the study of drugs that affect the nervous system.

These drugs include anesthetics (eliminate sensation), anticonvulsants (used to treat epilepsy), analgesics (relieve pain), and a variety of drugs that affect the autonomic nervous system." [3]

"Psychotropic drugs exert their effects by altering a synaptic event.

These alterations ultimately change the activity of a neurotransmitter. Some psychotropic drugs facilitate the effects of a neurotransmitter, and are called *agonistic*.

While other psychotropic drugs inhibit the effects of particular neurotransmitters, and are called *antagonistic*." [3]

That means that drugs change the synapse from what it is to another form of component.

"In depression, there is a imbalance of chemicals in the brain. Antidepressant medications balance the brain chemicals serotonin and norepinephrine." [4]

"In a depressed person's brain, the neurotransmitters, norepinephrine and serotonin are not produced in sufficient quantities.

Because of this lack, too few messages get transmitted between neurons, and the symptoms of depression occur." [4]

Without any regard for what causes neurotransmitters to be 'low' or 'high' drug therapy addresses the results instead of the causes.

Essentially, the synapse is a diode.

Since the architecture is made biologically the form of component that would be able to stop a pathway's signal from reversing would be a biological function of accepting the wave amplitude from the previous neuron transmitted by the axon and converting it into a transmission of chemicals.

Those chemicals cross the very close distance to the receptor of the neuron. The ones that make it are absorbed by the neuron, which transmits the translated value back into a wave amplitude.

It is exactly the same component (biologically constructed) as a diode component.

"A diode functions as the electronic version of a check valve. By restricting the direction of movement of charge carriers, it allows an electric current to flow in one direction, but blocks it in the opposite direction. " [1]

"A diode is the simplest possible semiconductor device. A diode allows current to flow in one direction but not the other. You may have seen turnstiles at a stadium or a subway station that let people go through in only one direction. A diode is a one-way turnstile for electrons." [5]

"When you put N-type and P-type silicon together as shown in this diagram, you get a very interesting phenomenon that gives a diode its unique properties." [5]

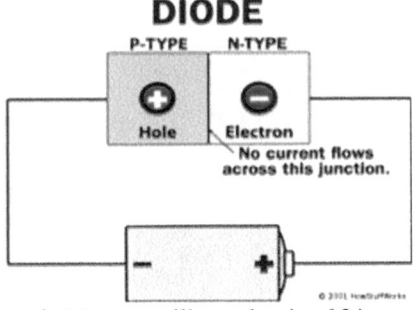

"Even though N-type silicon by itself is a conductor, and P-type silicon by itself is also a conductor, the combination shown in the diagram does not conduct any electricity.

The negative electrons in the N-type silicon get attracted to the positive terminal of the battery. The positive holes in the P-type silicon get attracted to the negative terminal of the battery. No current flows across the junction because the holes and the electrons are each moving in the wrong direction." [5]

"If you flip the battery around, the diode conducts electricity just fine. The free electrons in the N-type silicon are repelled by the negative terminal of the battery.

The holes in the P-type silicon are repelled by the positive terminal. At the junction between the N-type and P-type silicon, holes and free electrons meet. The electrons fill the holes. Those holes and free electrons cease to exist, and new holes and electrons spring up to take their place. The effect is that current flows through the junction. " [5]

"A device that blocks current in one direction while letting current flow in another direction is called a diode.

Diodes can be used in a number of ways.

For example, a device that uses batteries often contains a diode that protects the device if you insert the batteries backward. The diode simply blocks any current from leaving the battery if it is reversed -- this protects the sensitive electronics in the device." [5]

The result in the biological synapse is transmitters.

"Since we do not yet understand exactly how a synapse works, it remains to be determined whether apparently unreliable synapses might turn out to be reliable under the right conditions (perhaps analogously to how we found that neurons were much more reliable for some times of inputs than for others). This would be a very exciting finding." [6]

"Our research addresses a very basic question about how the nervous system works: is the brain essentially machine-like, using very precisely timed signals to process information, or is its operation essentially random and noisy?

Although this question has been around for many decades, it has not yet been answered for the cerebral cortex, the part of the brain that provides us with our highest mental abilities. Our experimental results show that cortical neurons are capable of extraordinary precision in the timing of signals that are used to communicate between them.

Although it is not yet known whether cortical neurons use this precision to communicate information, the measurements we have made provide strong incentive for looking more closely at possible temporal codes.

The answer will profoundly impact our theories of how the neural activity in the brain gives rise to complex animal and human behavior." [6]

The fact that the question, "is the brain essentially machine-like, using very precisely timed signals to process information, or is its operation essentially random and noisy?" [6], could be asked at all is indicative of visual dominance in evaluating a process and perpetuates the desire of most visual scientists to prefer random to logic as logic would imply responsibility.

"The implication of our evidence for a precise neural mechanism is that the exact timing of spikes may very well be used by neurons to send information to each other.

This suggests that researchers ought to pay careful attention

144

in future experiments to the exact patterns of activity occurring at very brief time scales.

Generally speaking, this makes for harder experiments than simply measuring and interpreting activity over a longer time scale. We will also need to consider more carefully theoretical models of exactly how the brain can make use of precise temporal information (the question of the "neural code").

The field of neuroscience is on the brink of appreciating the complexity and potential implications of temporal neural computation." [6]

Drugs to combat mental conditions usually work on those transmitters by changing the output (the transmitters) of the synapse's result based in timed and orchestrated clocking patterns to a representational value of the neuron receptor closer to the value the input receptor is giving for processing, artificially creating a memory that does not exist in reality.

MEMORIES:

When a person thinks of memory, having a rudimentary knowledge base of other forms of memory is it logical to deduce that if one memory is static then all memory is static?

It is logical to assume that a 'memory' is both the entire event or stimulus and the individual parts of that event or stimulus.

For the purposes of clarity this piece refers to 'memory' as the collection of parts while referring to a part of 'memory' as a wave amplitude.

How one becomes the other is fascinating.

Imagine your vision. Each eye contains rods and cones: regulating cells that when excited will permit the passage of brain dynamic system charge in direct proportion to the degree of excitation.

That means that the flow of charge is controlled by each individual receptor, each being a part of the whole of each eye, both eyes being a part of the whole of the entire event or memory.

Within the brain, that regulated charge is processed within its own pathway, like hundreds of thousands of sponges.

The sponge represents a wave frequency or pulse wavelet of one receptor. It travels within the axon to the next neuron, which is more like a controlled demolition derby, smashing one sponge into another, each giving off half of their contents.

The contents are then pushed out of the neuron in a new sponge. That is the computation part starting at the point of input receptor value meeting prior memory value:

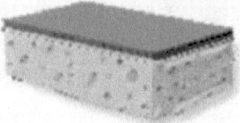

Empty Sponge = Wavelet With No Amplitude

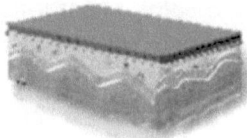

Sponge after the input receptor gives Amplitude

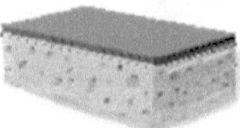

Many Memory Sponges meet

The Input Sponge, Which Results In

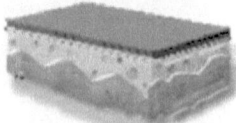

Many Comparative New Amplitude Sponges

The process of input amplitude compared to memory amplitude takes place in a pulse advance process where memory sponges are moving faster than input sponges.

That makes a 'ratio-enhanced' or exponential computation where each subsequent sponge receives a slightly less amplitude, setting up the concept of time. [8]

The input receptor's structure where all receptors within the 'device' are receiving a slightly out of focus stimulation provides a three dimensional aspect to the whole of the memory while giving each neuron wave amplitude pathway a single part of it. [9] [10]

With each comparison to another wave amplitude pulse the amplitude reduces. When it returns to input comparison it comes back in the same order it was made setting up a relationship to now and past time.

Each of those computations results in a synapse conversion and transmission.

If input values are not reduced due to a defect in the 'hardware' (you can still see, you can still hear, or if deaf or blind or both you still have the same form of remaining input value your brain has relied on for movement) then the input values are not a problem and the processing values are.

That means that once a wave amplitude has been established in memory from a normal input receptor pathway it will go down memory (both long and short-term) with time normally. A condition of 'down-string' or time-based complaint is not normally associated with processing values of depression; the 'shakes' or other 'software' or system based conditions and is not the same thing as the rectifier based synapse.

The effect of drugs either amplifies or reduces return feedback amplitudes which result in lower short-term wave amplitudes, less relation to long-term amplitudes and a general feeling of being 'balanced' or otherwise 'normal'.

The cycle, once set up of decreasing return amplitudes being fed by decreased new amplitudes leads to a form of vegetation-like responses and the need for more drugs which leads to the need for more drugs.

There is something illogical in taking your brain to the 'repair' shop to be treated by a person who pays attention only to results and symptoms and has not one clue as to what causes it all.

By treating the result of a brain function instead of the process using that brain function drugs have given hope to the bothered and hell to pay for using them.

There is a better solution.

Synchronize, not anesthetize.

Just as each neuron is pulsed in its specific role within the symphony of memory processing the whole made of those parts likewise are pulse or time-rate dependent.

The human brain is made of two main parts; long-term and short-term processing. Output, input, reduction of level processing rates and other ancillary yet important parts of the system are biologically based and not part of this treatise.

The whole of the brain is made up of two main parts; visual and aural pathways in both short and long-term memories. Other senses also retain memory, process comparisons and output to motion but not in the depth of visual and aural pathways.

After all, you either see or hear all the time you are awake and only smell something different from time to time and taste something different perhaps three times a day.

Those pathways are amplitude charge waves that use shared neurons defaulting to a perfectly orchestrated firing pattern in a healthy brain.

Physically tracing such a pathway would require far too intrusive measurement connections resulting in decoherence of the wave and elimination of the pathway's charge state rendering the measurement moot.

Short-term is either visual or aural dominant.

Long-term is either visual or aural dominant.

Dominance is determined physically, by which pathway's pulse rates most closely match a division of the input receptor's pulse rate.

It is a very fine range.

Each input receptor putting out wave amplitude sponges does so twice a second: quite a slow speed for the clarity of our vision and hearing.

Luckily our eyes are not like cameras. Each image from the eye is actually a series of images (snapshots) taken by pulse rate firing dependent distributed receptors all receiving data from the same focused lens.

The image only becomes a single dimensional image when it is assembled in brain processing in the same time frame.

When the image is complete, a 'recognition' will be made of past memory.

If the image is not complete (or the concept does not ring true) the missing parts are missing because they have either been compared to other things far more than the imagined memory or the rest of the story is contained in a pathway that is not in the same train car.

Memory is also like a train.

Where each sponge is moving from one train car to the next, as the train passes by the sponge is essentially standing still, touching each car as it passes and outputting a new sponge in the train car with each touch.

Those wave pulsed amplitudes represent a part of that memory. When they are compared to other parts of the memory and they are not doing so at the same pulse rate the sum of the event memory from the slower pathway will not equal the sum of the event memory in the normal or advanced memory pathway and the result will be:

Reduction of synapse transmission in that combined new memory.

If the reduced memory pathway is part of the whole that makes up the person's sense of self or self-awareness the person will be depressed.

If the reduced memory pathway is part of the whole that is the subservient pathway to the person's sense of self or self-awareness the person will be agitated.

Since synapses are involved in both conditions treating the same location in opposing ways addresses and masks the symptoms of the depression or hyperactivity instead of tuning the pathways to let the brain make more sense of itself.

Where short-term memory takes up the most room in the human brain, long-term works in the exact same way so our depiction here will show them equal in size for clarity of the exercise and the details of how they work differently will be left to the reference of chapter two".

If asked which form of thinking you employ you would respond at first with a slight disturbance from long-term memory not having contemplated or not having enough empirical evidence to support such a concept.

Depending upon the deductions already contained in your long-term memory over the concept of a 'form of thinking' versus a random symphony of not understood chemical reactions, the second response would indicate whether you are buying any of this paper at all.

Rest assured that whether you 'buy into' this paper or not does not change the fact that describing an event using the mechanism of this paper managed to predict not only your rejection of unknown things but also your degree of refusal.

Exercising short-term dominance over long-term reactionary memory is a wonderful experience.

If you had managed to react with a response your response would be one of four things:

1. Visually, because you see images in your head.

2. Aurally, because you hear words in your head.

3. Both, because you see and hear (other than dreaming) in your head.

4. Neither, because you either do not understand the question or are refusing to admit that one or the other is dominate.

The part of your brain that is responsible for your knowing which or none or both dominate sense is the short-term process running at an advanced pulsed rate over long-term process by a factor of 2. (900:1 instead of long-term's 30:1).

Long-term was that nagging rejection you felt to the most recently contradictive input you received. Long-term is also the culprit for the 'conscience' we seem to perceive from time to time.

Used in the currently in vogue interpretation of Freud 'conscience' is defined as "the part of the superego in psychoanalysis that transmits commands and admonitions to the ego." [7] It is also your past.

Many people do not 'perceive' a 'conscience' feeling or voice or fleeting and seemingly unrelated image. Those people are usually most susceptible to long-term influence and therefore are not aware of the separate nature of the messages.

For a person not dominated by or at least equal in short-term processing to long-term processing the normalness of long-term messages will not indicate a separate entity or source.

For a person dominated by or at least equal in short-term processing to long-term processing the normalness of long-term messages will indicate a separate entity or source.

Resulting personality characteristics can be based in essentially positive prior input or essentially negative prior input.

Positive prior input (environmental and self-determined)

would be viewed as the 'conscience' most people perceive as that 'inner-voice' (in females it is called intuition since it is normally visual) and would be accepted as a nudge to review a specific input event for further causes or actions based in potentials which are based in past experience.

Negative input (environmental and self-determined) would be viewed as the 'little devil' some people perceive as the 'demon' inside or in a lack of connectivity to the current event that experience would dictate otherwise and would be accepted as a nudge to review a specific input event for further causes or actions based in potentials which are based in past experience.

So, which are you, primarily?

VISUAL OR AURAL?

There are a few generalities that can be applied to a species starting with its primary sense and its primary level of processing (long-term to input).

Humans are at first an aural species. Hearing outputs to the most precise output actuators dominating all other outputs to that precise actuator type. The result is speech, language and communication.

The first external input processing center accessed by a living human being is hearing. It is the only 'sense' able to be activited by external input through the womb.

Humans, like most other species is split into two main parts, by gender:

Female is visual.

Male is aural.

That means that long-term memory is by default faster in visual for females and faster in aural in males.

The term 'faster' does not mean out of synch. It means corrupted synchronization and it is not at all very much difference.
An input receptor's arbitrary (just for this exercise) hypothetical value of allowing an amplitude of 10 to pass with a wave pulse into comparison with long-term memory values representing the same

151

time frame of reference processing at a 30:1 enhancement would mean in one second such input amplitude of 10 would be presented to 30 long-term memory amplitude pulses resulting in 30 new amplitude values, a mean sum of the two.

If the previous processing of wave amplitudes had not been set up yet (the brain has just started functioning) then the same base value (the internal body clock's base frequency amplitude, which is a non-zero value we shall represent here as .01) would create the first set of 30:1 parts of the ½ second whole.

$$\text{Amplitude } 1 = (((10-.01)/2)+.01) = 5.005$$
$$\text{Amplitude } 2 = (((10-.01)/2)+.01) = 5.005$$
$$\text{Amplitude } 3 = (((10-.01)/2)+.01) = 5.005$$
$$\text{Amplitude } 4 = (((10-.01)/2)+.01) = 5.005$$
$$\text{Amplitude } 5 = (((10-.01)/2)+.01) = 5.005$$
$$\text{Amplitude } 6 = (((10-.01)/2)+.01) = 5.005$$
$$\text{Amplitude } 7 = (((10-.01)/2)+.01) = 5.005$$
$$\text{Amplitude } 8 = (((10-.01)/2)+.01) = 5.005$$
$$\text{Amplitude } 9 = (((10-.01)/2)+.01) = 5.005$$
$$\text{Amplitude } 10 = (((10-.01)/2)+.01) = 5.005$$
$$\text{Amplitude } 11 = (((10-.01)/2)+.01) = 5.005$$
$$\text{Amplitude } 12 = (((10-.01)/2)+.01) = 5.005$$
$$\text{Amplitude } 13 = (((10-.01)/2)+.01) = 5.005$$
$$\text{Amplitude } 14 = (((10-.01)/2)+.01) = 5.005$$
$$\text{Amplitude } 15 = (((10-.01)/2)+.01) = 5.005$$
$$\text{Amplitude } 16 = (((10-.01)/2)+.01) = 5.005$$
$$\text{Amplitude } 17 = (((10-.01)/2)+.01) = 5.005$$
$$\text{Amplitude } 18 = (((10-.01)/2)+.01) = 5.005$$
$$\text{Amplitude } 19 = (((10-.01)/2)+.01) = 5.005$$
$$\text{Amplitude } 20 = (((10-.01)/2)+.01) = 5.005$$
$$\text{Amplitude } 21 = (((10-.01)/2)+.01) = 5.005$$
$$\text{Amplitude } 22 = (((10-.01)/2)+.01) = 5.005$$
$$\text{Amplitude } 23 = (((10-.01)/2)+.01) = 5.005$$
$$\text{Amplitude } 24 = (((10-.01)/2)+.01) = 5.005$$
$$\text{Amplitude } 25 = (((10-.01)/2)+.01) = 5.005$$
$$\text{Amplitude } 26 = (((10-.01)/2)+.01) = 5.005$$
$$\text{Amplitude } 27 = (((10-.01)/2)+.01) = 5.005$$
$$\text{Amplitude } 28 = (((10-.01)/2)+.01) = 5.005$$
$$\text{Amplitude } 29 = (((10-.01)/2)+.01) = 5.005$$
$$\text{Amplitude } 30 = (((10-.01)/2)+.01) = 5.005$$

This sets up an event memory whole with no sense of time.

5.005 | 5.005 | 5.005 | 5.005 | 5.005 | 5.005 | 5.005 | 5.005 | 5.005 | 5.005 | 5.
005 | 5.005 | 5.005 | 5.005 | 5.005 | 5.005 | 5.005 | 5.005 | 5.005 | 5.005 | 5.00
5 | 5.005 | 5.005 | 5.005 | 5.005 | 5.005 | 5.005 | 5.005 | 5.005 | 5.005 |

If there were no slight variance in a pathway's clock rate (if every clock rate division in the brain were perfect) then no sense of time would set it at the initial level of processing (it would set in later in short-term processing).

In a 'perfect' system the amplitude sequence of:

5.005 | 5.005 | 5.005 | 5.005 | 5.005 | 5.005 | 5.005 | 5.005 | 5.005 | 5.005 | 5.
005 | 5.005 | 5.005 | 5.005 | 5.005 | 5.005 | 5.005 | 5.005 | 5.005 | 5.005 | 5.00
5 | 5.005 | 5.005 | 5.005 | 5.005 | 5.005 | 5.005 | 5.005 | 5.005 | 5.005 |

Would match the same 10 input amplitude in whole but since we are not perfect let us examine a simple 1 pulse rate difference; (with the same pulsed whole passing the same input again but one less enhancement:

Amplitude 1 = $(((10-5.005)/2)+ 5.005) = 7.5025$
Amplitude 2 = $(((10-5.005)/2)+ 5.005) = 7.5025$
Amplitude 3 = $(((10-5.005)/2)+ 5.005) = 7.5025$
Amplitude 4 = $(((10-5.005)/2)+ 5.005) = 7.5025$
Amplitude 5 = $(((10-5.005)/2)+ 5.005) = 7.5025$
Amplitude 6 = $(((10-5.005)/2)+ 5.005) = 7.5025$
Amplitude 7 = $(((10-5.005)/2)+ 5.005) = 7.5025$
Amplitude 8 = $(((10-5.005)/2)+ 5.005) = 7.5025$
Amplitude 9 = $(((10-5.005)/2)+ 5.005) = 7.5025$
Amplitude 10 = $(((10-5.005)/2)+ 5.005) = 7.5025$
Amplitude 11 = $(((10-5.005)/2)+ 5.005) = 7.5025$
Amplitude 12 = $(((10-5.005)/2)+ 5.005) = 7.5025$
Amplitude 13 = $(((10-5.005)/2)+ 5.005) = 7.5025$
Amplitude 14 = $(((10-5.005)/2)+ 5.005) = 7.5025$
Amplitude 15 = $(((10-5.005)/2)+ 5.005) = 7.5025$
Amplitude 16 = $(((10-5.005)/2)+ 5.005) = 7.5025$
Amplitude 17 = $(((10-5.005)/2)+ 5.005) = 7.5025$

Amplitude 18 = (((10-5.005)/2)+ 5.005) = 7.5025
Amplitude 19 = (((10-5.005)/2)+ 5.005) = 7.5025
Amplitude 20 = (((10-5.005)/2)+ 5.005) = 7.5025
Amplitude 21 = (((10-5.005)/2)+ 5.005) = 7.5025
Amplitude 22 = (((10-5.005)/2)+ 5.005) = 7.5025
Amplitude 23 = (((10-5.005)/2)+ 5.005) = 7.5025
Amplitude 24 = (((10-5.005)/2)+ 5.005) = 7.5025
Amplitude 25 = (((10-5.005)/2)+ 5.005) = 7.5025
Amplitude 26 = (((10-5.005)/2)+ 5.005) = 7.5025
Amplitude 27 = (((10-5.005)/2)+ 5.005) = 7.5025
Amplitude 28 = (((10-5.005)/2)+ 5.005) = 7.5025
Amplitude 29 = (((10-5.005)/2)+ 5.005) = 7.5025

(first of next event) Amplitude 1 = (((10-.01)/2)+ .01) = 5.005

Resulting in:

7.5025|7.5025|7.5025|7.5025|7.5025|7.5025|7.5025|7.5025|7.5025|
7.5025|7.5025|7.5025|7.5025|7.5025|7.5025|7.5025|7.5025|7.5025|
7.5025|7.5025|7.5025|7.5025|7.5025|7.5025|7.5025|7.5025|7.5025|
7.5025|7.5025|

Making the next event

5.005|7.5025|7.5025|7.5025|7.5025|7.5025|7.5025|7.505|7.5025|7.5
025|7.5025|7.5025|7.5025|7.5025|7.5025|7.5025|7.5025|7.5025|7.5
025|7.5025|7.5025|7.5025|7.5025|7.5025|7.5025|7.5025|7.5025|7.5
025|7.5025|

When this difference is built up over many events a sense of the end of an event being smaller than the beginning of the event sets up and the concept of time is evident.

The problem with improperly timed sequences is that the time notion remains but the relevance of each group of memory wave amplitudes does not.

Since input is still making things relative to environment based on 2 cycles per second the divisions of those enhanced processed wave amplitudes start to develop a time concept that is distributed over additional parts of seconds and turn into additional parts of minutes and a habit can be formed from simply facing an event with an inopportune pathway.

That might be viewed as an inherited trait or propensity for some form of abusive or non-constructive behavior.

These divisions are set genetically. They result in a 'system-whole' representing the entire processing pathway combination of a specific sense.

If vision is that sense and it is closest to being normal to input it will be the dominant long-term process. If aural is that sense and it is closest to being normal to input it will be the dominant long-term process.

So which are you, aural or visual?

You would have to know how you think short-term (consciously) versus long-term (sub-consciously) to determine which is dominant in either and therefore which is dominant in the entire brain.

If you are aural dominant long-term and visually balanced short-term your memories of past experiences will be conceptual and it will take a bit consciously to summon a visual representation of that memory event.

If you are visually dominant long-term and aurally balanced short-term your memories of past experiences will be visual perception and it will take a bit consciously to summon an aural conceptual relationship to a current event.

Which short-term process is dominant will determine how easy the conscious evaluation to:

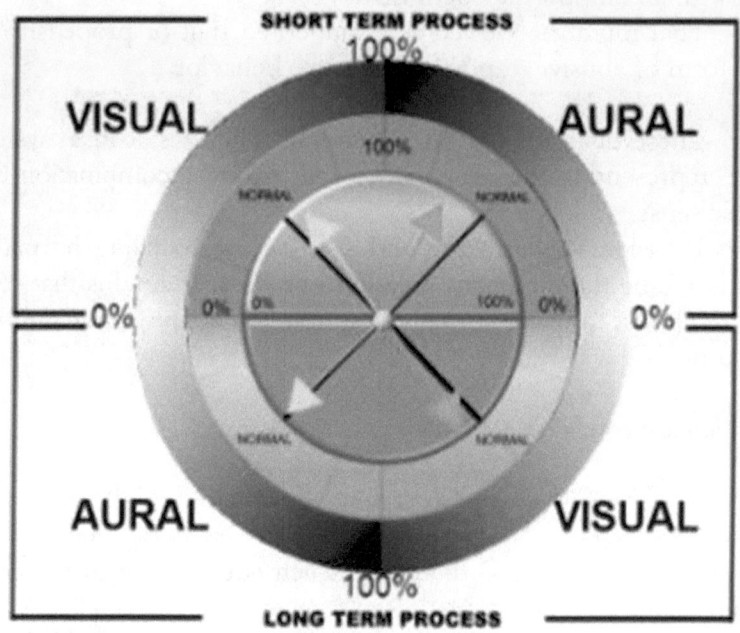

The chart has an inner circle with its legend in the middle circle and the overall brain result as the outer circle.

It shows Freda's short-term visual process as normal (slightly advanced equal to the long-term advanced female visual process) and short-aural process as advanced over normal aural long-term.

The imbalance is evident in aural where long and short-term are not equal. That means concepts of events readily recallable in visual long-term are not equal to those same events recalled in aural long term.

So parts of events that cause similar visual long-term recall are judged by the aural short-term process as normal and the cycle of staying in the same emotional mess repeats.

The solution to equal comprehension of events in their correct time frame for Freda is to increase the value of aural event wave amplitudes by consciously forcing a conceptual relationship based in 'now' with past visual relationships and building a new correlation of them based in conscious 'now' instead of sub-conscious 'then'.

The more Freda exercises using her 'softer' conscious process the more the values represented by wave amplitudes in her long-term aurally dominated memory will increase thereby accomplishing the same result a drug accomplishes by increasing synapse intensity.

The drugs are no longer needed and the result is the same only long-lasting and without side effects unless you can consider far less compulsiveness and far more reasoning to be a side effect.

Fred has only begun to understand how his brain does what it does and with time will be able to accomplish the same result.

This is what Fred's dominance chart looks like:

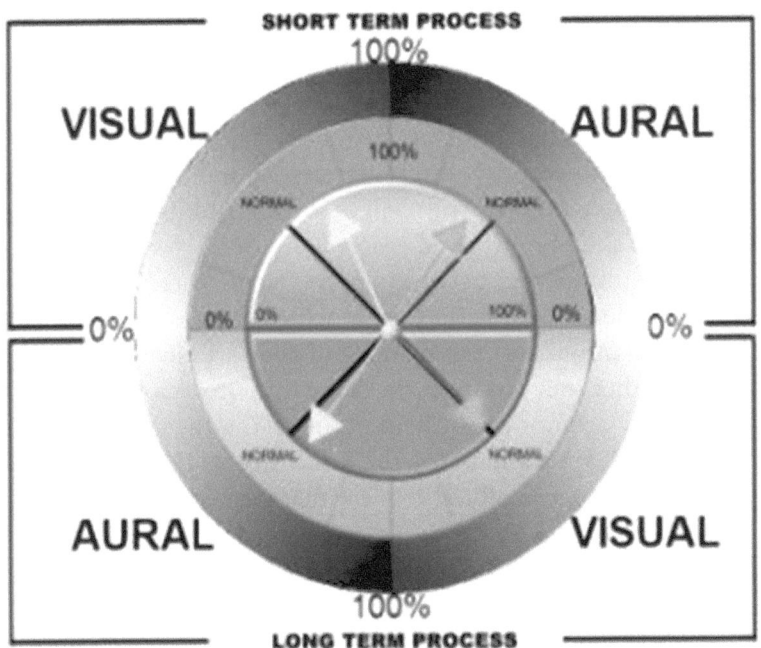

It shows Fred's short-term aural process as normal (slightly advanced equal to the long-term advanced male visual process) and short-visual process as advanced over normal visual long-term.

The imbalance is evident in visual where long and short-term are not equal. That means concepts of events readily recallable in aural long-term are not equal to those same events recalled in visual long term.

So parts of events that cause similar aural long-term recall are judged by the visual short-term process as normal and the cycle of staying in the same emotional mess repeats.

The solution to equal comprehension of events in their correct time frame for Fred is to increase the value of visual event wave amplitudes by consciously forcing a conceptual relationship based in 'now' with past aural relationships and building a new correlation of them based in conscious 'now' instead of sub-conscious 'then'.

The more Fred exercises using his 'softer' conscious process the more the values represented by wave amplitudes in his long-term visually dominated memory will increase thereby accomplishing the same result a drug accomplishes by increasing synapse intensity.

The drugs will be no longer needed and the result will be the same only long-lasting and without side effects unless you can consider far less compulsiveness and far more reasoning to be a side effect.

Since Fred has only just begun to understand how his brain works his path to recovery from depression and his overcoming the genetic imbalance he lives with will be a long and intense process but luckily he can do it on his own.

The difference between 'now' and 'then' in short-term processing is dependent upon how much awareness one has of 'now'.

Wave amplitudes can match and support recall of an image or concept whether they are from the same time frame or not.

If they are not of the same time frame the reaction a person has may not be appropriate for the 'now' event. Just because the car was red does not mean all red cars are that past car.

DEALING WITH A SUDDEN CHANGE:

If you have taken medication over time that is conducive to needing more to feel the same effect (habit forming drug) then you have experienced the overwhelming and seemingly uncontrollable urges, movements and antsy feeling brought about by stopping a synapse effecting drug.

The drug has either raised synapse transmitters to match levels a timed orchestrated brain would have and when removed has thrown you into a nearly uncontrollable mood change or depression cycle or the drug has lowered synapse transmitters to match levels of a timed non-orchestrated brain would have and when removed has thrown you into nearly uncontrollable mood change or depression cycle the outcome is the same.

You need the synapse levels to be at the range the drugs gave you in order to feel balanced but your brain is not genetically built to make that happen and you have lived for years with the unbalanced wave amplitudes telling you reality is something else.

That 'something else' can be the result of no drug intervention when timing is off or the result of drug intervention by putting the timing back where the genetics said it was supposed to be.

Either way, you can control a degree of the timing difference by correcting the imbalance of visual and aural dominance in long-term memory with short-term memory evaluation.

Read this chapter as many times as it takes for you to acquire an awareness of how the brain works. Read chapter two again, for further detail if necessary.

You will know when you are ready to start taking control of your brain when you know how the brain works and therefore can use that mechanism to make it work right.

If Fred's condition was the result of drug withdrawal the solution is the same. Concentrate on increasing the long-term process wave value of the dominate long-term process by employing the dominant short term process and it will increase softer short-term in the same process. If done enough it will equal short term process speeds of the dominant short-term sense and you will feel balanced again.

If Freda's condition was the result of drug withdrawal the solution is the same.
Concentrate on increasing the long-term process wave value of the dominant long-term process by employing the dominant short-term process and it will increase softer short-term in the same process. If done enough it will equal short-term process speeds of the dominant short-term sense and you will feel balanced again.

If you are an aurally long-term dominant male with aurally short-term dominant processing and you are withdrawing from a synapse enhancing drug use your aural short term concentration to focus on long term visual events, recall and describe those events and connect them to 'now' in the sense that they represent 'then'.

One exercise to make this easier to occur is to visualize words as you speak them.

See the word in your head as you speak the word. Make a practice of it and you will connect visual long-term to aural short-term and thereby increase visual long-term process values (replacing the lost increase from the drugs).

An even easier form of exercise for an aurally dominant short-term male is to listen to music and watch a small light pulse to the music.

Concentrate on the music and determine what concept the music conveys then force yourself to 'see' an image of that concept in your brain.

The closer to 120 beats per minute the music is the more tuning your brain will receive as long as the light pulses with each beat or every other beat as that nearly matches the division ratios of your brain.

For a visually dominant short-term male who is aurally dominant long-term the word visualization exercise would be reverse but the music exercise would be the same.

Tune the feedback from visual to match the clock rates performed by aural or tune the aural to match the clock rates performed by visual.

Music can sooth the savage beast and even the not so savage one as well.

ABOUT MOTION:

Every input has an output to motion.

Not every output to motion is strong enough to make a difference in the final movement. Normally visual processing is the strongest output to motion.

If you suffer from jitters, shakes or seemingly uncontrollable movements there is a simple way to begin to stop them.

You are suffering from such interruptions because long-term memory has a greater value than short term memory.

To solve it, do not make any movement unless you consciously decide to move. Watch every move take place by focusing on the limb moving and looking at it.

This will help time the motion with the output (which will cut back on the jitters) and synchronize the input visual stimulus with the output visual based motion.

Try it. If your hand shakes when you pick up a glass of softdrink do not pick up the glass unless you have a: decided to pick up the glass and b: watched your arm reach for the glass, grab the glass and raise it up.

The more you concentrate on the arm moving the less shakes you will feel.

Keep it up and the shakes based in dominant long-term memory will no longer be a problem for you since your conscious level process will have taken control.

The following graphic is for your own chart.

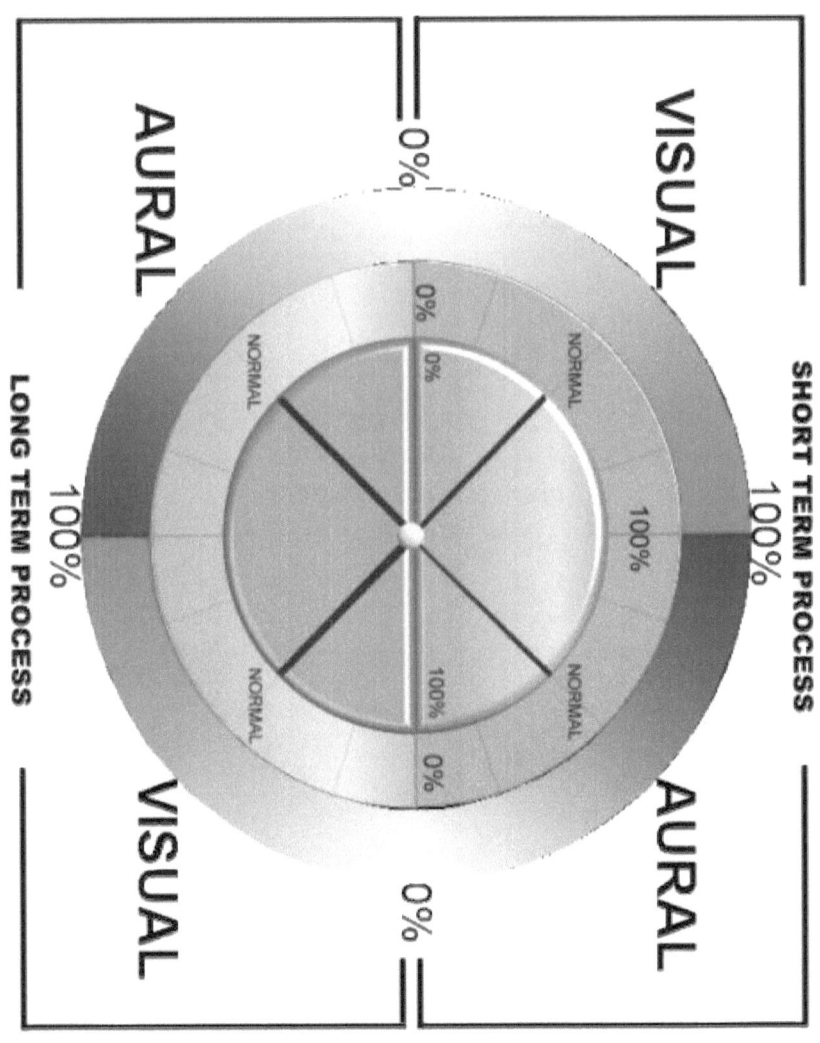

Chapter Nine
Evolution of Knowledge

Fixation & Dependency

Have you ever heard of William Thomson, Lord Kelvin? Thomson was a brilliant mathematician, who can take credit, for being the spark that set Maxwell into creating a new theory of electromagnetism, even though he disagreed with it. [1]

He observed the Joule-'Thomson' effect, in 1852.

"The author of the biography of Thomson, puts forward the view that during the first half of Thomson 's career he seemed incapable of being wrong while during the second half of his career he seemed incapable of being right.

This seems too extreme a view but Thomson's refusal to accept atoms, his opposition to Darwin's theories, his incorrect speculations as to the age of the Earth and the Sun, and his opposition to Rutherford's ideas of radioactivity, certainly put him on the losing side of many arguments later in his career." [1]

"Radio has no future. Heavier-than-air flying machines are impossible. X-rays will prove to be a hoax." -- *William Thomson, Lord Kelvin, British scientist, 1899.[2]*

As with all other things, knowledge starts off slowly, progresses where each element within it breeds other elements and finally reaches a stage of balance where knowledge becomes a thing of its own and defends its right to exist against all odds.

In my EnticyPressed forum board signature, I quote 'Fred': "The true measure of a man's character, is his level of tolerance, for what he does not understand."

Another quote I find interesting, although irrational is posted at CogNews.Com."Science is a willingness to accept facts even when they are opposed to wishes." -- B. F. Skinner

Sometimes we can become so caught up in our knowledge that we say things without thinking them through: "I see no good reasons why the views given in this volume should shock the religious sensibilities of anyone." -- *Charles Darwin, The Origin Of Species, 1869. [2]*

A wish is based in knowledge.

It is the desire for fulfillment of a knowledge that is not the real thing, knowledge.

If science were to not accept a fact, due to a wish, then it would not be science it would be old country style fire and brimstone religion. .

The more a person acquires a specific knowledge, about a specific thing, the more that person will come to believe the knowledge, regardless of what the thing has to say about it.

If that knowledge is a brand new thing and exciting and consuming then the manner in which the person thinks will have a great deal to do with whether they display character in inquiring about that knowledge, or if they display fixation and dependency.

Fixation is focus without intent.

Dependency is focus without fulfillment.

Both intent and fulfillment can only be injected into a person's thoughts by the short-term 'mind'. The long-term memory is far too busy supporting the past.

That is all long-term memory is. It is a record, in order of processing, of all things experienced by the senses.

In humans, the short-term process, also a memory, but far shorter in duration, yet far longer in pathway depth, due to its vastly increased processing speed, feeds the long-term memory. It is how we remember who we are and that 'now' is not the same thing as 'then'.

Long-term memory lives in the past. Short-term memory lives in the present. It is far too short (7 seconds) to establish much other than an awareness of 'now'. The dichotomy of the existence of both is our perception of the world.

Just like all other dichotomies, they do not need to be of equal intensity to remain opposing.

Long-term memory, left alone, without interference of short-term processing will become the focus of the person.

If that focus is based in a long-held belief, or a mass knowledge base of inter-related figures or images, long-term, without much control of short-term, will result in the past doing the talking.

And it will defend its condition equal to the degree in which that condition is rooted.

A number of years ago, while attempting to kick-start Neutronics Technologies Corporation I took what I was aware of and headed into the 'field' to make my case.

You are welcome to view the list of many postings to Usenet from 1996 on, by searching for my name.

Some are good, some are frustration and some are ridiculous

responses rooted in insecurity.

Insecurity is focus with intent, but without fulfillment.

You will find listings on the brain, the Neutronics Dynamic System and weather warning systems.

At the time, I was not capable of writing material, readable by visually thinking persons: and it showed.

I also made a mistake in addressing the technology to scientists, or at least to want-to-be scientists posting in science groups under fictitious names: too much knowledge of the past, too little room for the future.

As if this material is somehow hidden and unavailable for anyone to review at leisure, it has become a crutch for today's believers in the past knowledge they live by.

Fixation on attacking a new thing, means dependency is on the old thing.

This time, 10 years after this technology first actually worked, 7 years to the month (July) after Little Ricci was built and wandered the yard on its own to prove the technology, the articles posted at EnticyPress have come under fire, not for the discussion of the brain, but for the Neutronics Dynamic System, and this time, instead of from Usenet retirees, from much younger memories.

That is to be expected.

Knowledge grows.

From the first knowledge of the use of the 'thumb', to today's preponderance with guessing enough to hit the jackpot by chance, knowledge has grown, faster and faster.

Unlike past attempts to simply monkey-fie a speaker, today's younger, more recently enlightened knowledge holders, defend their knowledge with far more vigor.

The reader is welcome to drop in, and see the manner in which, the attacks, are taking place these days.

I am undoubtedly unaware of most of the locations online, where ignorance defends its right to exist, but I am aware of one, otherwise fine and very useful discussion place that has become the playground for the knowledge kings.

By clicking on this link you will find the first discussion thread, independently started in 8 years that I am aware of, regarding the Neutronics Dynamic System.

The only reason a person would become so fixated and so

intent on doing something in his own field is due to the degree of knowledge that needs defending or the degree of un-fulfillment that needs filling.

There could be other threads 'out there' of a similar nature but I doubt it. Healthy adults do not normally make it a habit to wile their lives away in proving how big they think they are, or how much they think they know, or how bad they think they can be.

The more knowledge becomes trusted, the more knowledge it stops.

Thomson was a victim of his knowledge. Some people are victims of their knowledge every single day.

Knowledge is long-term memory. It is the first output to motion and as such, if not controlled by the short-term 'mind' will do pretty much what it wants to, based on what it has always done before. It cannot create. It can only repeat.

If it is given input from short-term processing that smacks as not in previous memory, it will, every single time, win, unless the person truly contemplates.

Contemplation is where innovation takes place.

Too much comfort in a set of agreed upon past knowledge, without contemplation, (the very thing that training seeks to eliminate) results in error.

Thomas Watson knew everything there was to know about computers in 1943 when he exclaimed, "I think there is a world market for maybe five computers." [2]

In 1977, Ken Olsen, Chairman of Digital Equipment Corporation proclaimed, "There is no reason anyone would want a computer in their home."

Innovators had become comfortable with the knowledge that made them comfortable, and out came the infamous words:

"The concept is interesting and well-formed, but in order to earn better than a 'C', the idea must be feasible." -- *A Yale University management professor in response to Fred Smith's paper proposing reliable overnight delivery service. (Smith went on to found Federal Express Corp.)[2]*

"So we went to Atari and said, 'Hey, we've got this amazing thing, even built with some of your parts, and what do you think about funding us? Or we'll give it to you. We just want to do it. Pay our salary, we'll come work for you.' And they said, 'No.' So then we went to Hewlett-Packard, and they said, 'Hey, we don't need you. You haven't got through college yet.'" -- *Apple Computer Inc. founder*

Steve Jobs on attempts to get Atari and HP interested in his and Steve Wozniak's personal computer. [2]

College teaches students to remember the thoughts of others.

Sadly, far too many live those thoughts, support the work of others, and base their potential on the past's of others.

It need not be so.

(Depending upon the time duration of the causes of the depression, and the degree of removal from 'now'), the depressed person will find treatment both a great potential and a possible bad thing.

Any condition one finds one to be 'in' becomes defendable as long as short-term processing is not having a say in the thoughts.

Knowledge has evolved, just like everything else has evolved: slowly, to faster, to more efficient. The only problem with knowledge being more efficient is, it has to remain the same knowledge.]

I was told once, a long time ago, by Professor Aaron Sloman of The University of Birmingham, UK that I would some day, figure out how to say what I was saying.

Thank you Aaron, for the spark.

Chapter Ten
Neuroscience Computer Repair 101

Notice that just about all neuroscience press releases contain the word 'may'. It is a nice month (if capitalized) or a trigger word to denote: "speculation". So why do they guess and include the 'guess' as 'speculation' deep in the release?

So you won't see it.

After all, if you read those press releases with the word 'may' as the title, you would never believe it and funding would never increase, so it is buried deep in the text.

What if the method of neuroscience deduction and evaluation was applied to a computer? The neuroscience computer repair manual 'may' look something like this:

Neuroscience Computer Repair Manual:

"Computers and brains both need energy. Plug your computer into the wall, push a button, and it will get the power it needs to run. Pull the plug and it will shut down. Your brain operates in a different way. It gets its energy in the form of glucose from the food you eat. Your diet also provides essential materials, such as vitamins and minerals, for proper brain function. Unlike a computer, your brain has no off switch. Even when you are asleep, your brain is active. " [3]

(Let us ignore the 'off' switch comment for now, but we all know what 'dead' means.)

"The delicate contents inside your computer are protected by a hard cover. Your skull provides a similar function for your brain.

Nevertheless, the external and internal components of computers and brains are all susceptible to damage.

If you drop your computer, infect it with a virus, or leave it on during a huge power surge, your precious machine will likely be on its way to the repair shop.

When damaged parts are replaced or the virus-caused damage is removed, your computer should be as good as new.

Unfortunately, brains are not as easy to repair.

They are fragile and there are no replacement parts to fix damaged brain tissue. However, hope is on the horizon for people with brain damage and neurological disorders as scientists investigate ways to transplant nerve cells and repair injured brains." [3]

The parts of the machine:

1: The monitor
2: The processor
3: The printer
4: The keyboard
5: The mouse
6: The modem
7: The scanner

Altogether, we shall refer to these parts as 'the computer'.

The 'computer' works by an as yet, unknown process, that takes place in its 'brain'.

The 'brain' of the computer is located in the processor box.

We can tell what the processing box is doing by observing the 'monitor', the 'printer', and sometimes the 'modem'.

We can tell what the processor is processing by observing the 'keyboard', the 'mouse' and the 'scanner'.

Dealing With Problems of the Input:

If the 'scanner' is not 'scanning' we 'may' take a reading of the 'scanner' connection and if that reading is low, compared to normal 'scanner' readings we can inject an 'amplifier' to increase its levels and the 'scanner' will 'scan' better. It is pretty much like putting glasses on your eyes.

If the 'keyboard' is not 'keyboarding' we 'may' take a reading of the 'keyboard' connection and if that is reading low, compared to normal 'keyboarding' readings we can inject an 'amplifier' to increase its levels so the 'keyboard' will 'keyboard' better. It is pretty much like wearing a hearing aid.

If the 'mouse' is not 'mousing' we 'may' take a reading of the 'mouse' connection and if that reading is low, compared to normal 'mouse' readings we can inject an 'amplifier' to increase its levels and the 'mouse' will 'mouse' better. It is pretty much like yelling real loud at the computer to get its attention in a different area of focus.

If the 'scanner' is not 'scanning' we 'may' take a reading of the 'scanner' connection and if that reading is high, compared to normal 'scanner' readings we can inject a 'resistor' to decrease its levels and the 'scanner' will 'scan' better. It is pretty much like putting sunglasses on your eyes.

If the 'keyboard' is not 'keyboarding' we 'may' take a reading of the 'keyboard' connection and if that is reading high, compared to normal 'keyboarding' readings we can inject a 'resistor' to decrease its levels so the 'keyboard' will 'keyboard' better. It is pretty much like wearing earmuffs.

If the 'mouse' is not 'mousing' we 'may' take a reading of the 'mouse' connection and if that reading is high, compared to normal 'mouse' readings we can inject a 'resistor' to decrease its levels and the 'mouse' will 'mouse' better. It is pretty much like numbing the computer to stop its attention in a different area of focus.

We will use 'high-tech' observational equipment to test the levels. We will look at the results on the monitor. And we will employ the new functional magnetic imaging machinery to scan the processor to see where all of this stuff is happening. [Helpful safety hint: fMRI's fry computer processors.]

We will study 'dead' brains for a while until we can manage to get a fMRI to look inside the 'live' brain. (OK, this is stretching it a bit as a computer scanned in an fMRI would make a nice boat anchor.)

We do not have that problem in 'our' computer, because 'our' computer uses 'chemicals'. Right.

"Although computers and brains are powered by different types of energy, they both use electrical signals to transmit information. Computers send electrical signals through wires to control devices. (Your brain also sends electrical signals, but it sends them through nerve cells, called neurons. Signals in neurons transfer information to other neurons and control glands, organs, or muscles." [3]

So, we can justify a 'different kind of energy' by saying it is 'electrical signals' (which would decohere in a fMRI. Try it, go ahead, put a working radio in an fMRI and see what happens, but then again a radio is not 'chemical'. Right.)

Observational Techniques:

Watch the 'monitor' part of the 'computer'. It will show you everything that is coming out of the 'computer' and is the easiest way

171

to diagnose an error.

And if the monitor is not showing things 'properly' we can always control it as well. After all, it is chemicals.

If the monitor is not showing 'blue' very well, we can inject an 'amplifier' and increase the transmission of signals to the monitor from the processor and make the blue look right.

If the monitor is showing too much 'blue', we can inject a 'resistor' and decrease the transmission of signals to the monitor from the processor and make the blue look right. Right.

Most importantly we must remember that we do not know how the processor works. We have some idea of where things happen in the processor so we can make assumptions on those things based on where they happen.

Of course, the important things don't happen in specific regions or 'centers' so in such case as evaluating anything having to do with the processor always inject 'may' or 'perhaps' or 'could' in your evaluation results.

According to Searle the computer does not process information, it processes something else so we can't find the specific information like we would if we were to access a library so even though we can possibly replicate the process of things the computer does we of course, can't figure out what that really is.

And by all means, protect the knowledge we have by rejecting everything that possibly leads one to believe anything we think is in anyway, wrong.

Dealing With Problems of The Output:

In the case of the 'monitor': see above.

In the case of the 'printer': If the signals sent to the printer are not normal we can inject an amplifier in the line and increase them to the correct output. If they are low we can decrease them by injecting a resistor.

In the case of the 'modem': If the signals sent to the modem are not normal we can inject an amplifier in the line and increase them to the correct output. If they are low we can decrease them by injecting a resistor.

Remember, the brain works with electrical signals and that means it is just like a computer, but we know that the brain is not really a computer because it does things that are not really binary and does them with chemicals, but we know binary works in computers so let us continue to evaluate the processor by looking at its external parts and what those external parts do based in how we understand computers.

If the computer stops working, reboot it. If the brain stops working, bury it.

End of Manual

Give It A Break

Why in the world do they concentrate on the chemicals when the chemicals change? [1], [4], [5] Because they change, it means they should be changed artificially, without any regard whatsoever to what the brain does with them? [4] [1]

The most recent study shows chemicals start off one way or another, after that, the brain controls the chemicals. [5] [6]

Brain changes, change chemicals. Like the case of 'Freda': Chemicals were changed three times in a year. One that 'may' work replaced each one that didn't work. Getting off the chemicals only worked when the brain worked through understanding. Guess work, works like a shade-tree mechanic.

Of course the right nutrients are needed to make the tissue work at all so it perhaps, maybe, could, possibly, may, potentially lead to ground breaking discoveries of our research and as long as we keep the knowledge of how it really works, unknown, no one will catch on and we can continue work on maybe, could, possibly science.

Chemicals are then the focus, and we change chemicals regardless of why the brain changes them.

We disregard what the brain does with chemicals and concentrate on doing the chemicals ourselves.

Of course, the more we do that, the more dependent chemical users become to our chemical solutions, and the more we continue doing them. [7]

We'll buy up all the time on television so our patients even tell their doctors what to prescribe. But side effects 'may' cause (insert anything remotely far worse than the condition is).

Sometimes, researchers do not follow this manual and they

put things out that support the actual dynamic process of the brain. [2] [5] [6]

But who's listening? [7]

Chapter Eleven
Intelligence: What It Is,
How It Works

Ever since humans first knew of their own existence, the wonder of intellect has guided quests for knowledge.

The attempt to gain the knowledge of intellect, has led to logical and well-meaning interpretations, of what intellect does, where it comes from, and what it actually is.

This is the explanation of what Intelligence is and how it works and what causes it to emerge and in whom or what.

Intelligence Was:

The oldest recorded societies placed a different physical location to intellect than we know today.

There was a reason the heart was considered the source of intellect.

Most people, to this day, live their lives completely unaware of the central location of thought. Those people live in reactions, based in previous reactions.

When the heart was considered the center of intellect, the human species was still in its very low (and long wavelength) frequency of knowledge acquisition.

Knowledge was purely acquired through repetition, of very restricted environments.

Later, that exact process would emerge, as the regime of apprenticeship: the first teaching experiment.

As knowledge grew, first in what was considered outlandish, and often-chastised blasphemy, it increased the curiosity of those learned; which, increased the knowledge, which increased the curiosity.

One cannot hope to be curious, if one does not know that curiosity is the desire of knowing. One first has to know, that knowing is possible.

The level of knowledge required to first seek a difference in pro-action, versus reaction, came much later in the evolution of our general species, and its many variations.

When it did, creativity was born.

Creativity is not painting crude renditions of that night's meal on the tribe's cave wall.

It is using the past to seek the future, instead of using the past to guide the future.

How did the moment of consciousness happen? It is the first question, without knowing about any concept of 'consciousness', which sparked the first speculation about us and about where we came from.

With little knowledge to employ, and just as little self-awareness, the humans of ancient tribes sought out answers to what made them able to respond to danger, in a manner not indicative of the situation.

They pondered, and those who pondered most, became scholars, and were admired by others, as the source of knowledge.

The first scholars attributed the self to a mystical unknown, as they knew enough to be curious, but not enough to know.

One attribute of being aware of existence is the ability to be aware of a special existence, even if it comes from not knowing what it is yet knowing how to make it interesting.

With less and less repetition, in the lives of humans growing with more knowledge (even if it was as rudimentary as farming and boat building) the manner in which to retain the interest and therefore the focus, is repetition.

Then, the most curious became the second generation of scholars who knew enough to question the repetition and the mystical unknowns, as well as enough to not threaten them.

Those scholars contemplated sources, or causes, of what they observed and they attributed them to the mystical symbols of the day, as they still had not reached a level of knowledge, whereby knowing they knew, was possible.

Over more time, and in ever increasing magnitude, the level of knowledge exploded across the globe, as one new item of knowledge, led to many connections with other older items of knowledge, making many newer items of knowledge.

It is the process of an exponential system at work.

But knowledge is not intelligence.

Knowledge is the collection of experience, not the process of collecting experience.

That process is, intelligence, just as military organizations use the term.

In order to understand human intelligence, which is pro-active intelligence, one must understand: reactive intelligence derived from static response.

No matter how much processing the fastest super-computer in the world has it is still a simple static response.

What goes in comes out as it has been instructed.

The only difference between a binary computer and a mechanical counting device is the level of instruction.

The level of instruction is compounded by the level of processing, to provide the illusion of intelligence: if one does not know what it is.

Singularities:

It seems that no matter how much knowledge the species acquires, the search is still for that one thing that is the cause.

If you leave this piece with only one recallable concept, let it be this one:

Everything in our physically existing, solid matter Universe, is made up of two things, not one thing.

Intelligence is not a single thing.

The science of Artificial Intelligence defines 'intelligence' as either the dictionary's definition: "the ability to learn or understand or to deal with new or trying situations" [1] or the definition of each and every single AI researcher, based in their specific perspective and knowledgebase.

This unknown has led to every manner of misinterpretation.

At least this time there is enough knowledge to comprehend the reality of intelligence; no matter how defensive each and every AI researcher might be over its not agreeing with their perspective's deductions.

Computers today take in data, represented by a system of 1's and 0's that is a basic and simple logic mechanism.

By combining 1's and 0's faster and faster the illusion of reality emerges. The problem arises, when those who employ that illusion, believe it.

The logic of 1's and 0's reduces to combinations of 1's and 0's.

In executing that logic those combinations include either one.

1 or 0, 0 or 1, interpreted 1 and 0, 1 and 1, 0 and 0.

From these potentials, in order, presented faster and faster, an innumerable amount of results can occur.

What the combination cannot include is a combination of both.

Since 0 has no value combining it with 1 results in 1.

Why then, are we computing with a system that includes nothing and assigns it a useful value?

Natural systems, like that responsible for intelligence do not include 0. 0 means dead. There is no degree of dead. Dead is dead.

Contemplating the application of such a half dead, half live system itself is ludicrous to employ as any reference to a naturally emerging function. If such a process were involved in our observable, solid matter Universe there would be no solid matter Universe.

All causes are made up of two causes. Every cause is itself a result.

The one aspect of natural system today's computing logic employs correctly is the enhancement of speed.

Since the enhancement of speed is attributed to a static process it must be very, very fast to appear static. Less speed and the static process appears unstable and less 'real'.

Intelligence Is:

Intelligence starts at its first level through the enhancement of speed.

From the lowest brain form to the highest, this increase in speed is paramount to intelligence emerging.

But the enhancement is not static.

It starts with input.

Each input receptor of the brain samples its environment, through its specific perspective, afforded by its location and its degree of focus, two times a second.

Two times a second is a very slow process but since it is the lowest or first level of process for the emerging intelligence it must be as low as possible.

The sample is sent to the brain from the receptor to meet previous inputs from the same receptor.

A computer would stop right there and apply rules and commands and instructions for how to deal with the input and spit out a result: a static result. It may reach into prior instructions contained in its memory but it is not accessing the memory it is employing it.

179

Brains accept that input at two samples per second and present it to a memory process that is running faster.

Starting off at 1:1 in primitive evolutionary life forms with brains, the power of today's super-computers no matter how many parallel processes they execute, and increasing up to a mere 60 samples per second the brain's basic evolution emerged in the ability to output motion faster than input would statically dictate.

If a modern computer could process its data at twice the degree of input based upon previous input it would emerge with reactive intelligence and that is the process Artificial Intelligence has been trying to create.

But one does not duplicate a cause by making one of its results.

And one does not make an intelligent machine out of numerous modules of specific results.

The first level of intelligence is degree of advancement over input.

It is the same as the second level of intelligence, the human pro-active intelligence, but let us continue with level one:

The brain. It is the result of the creature's brain.

Cats and dogs as well as all other forms of creatures with brains possess reactive intelligence.

Memory is compared to input with output much faster than input would result in alone.

Motion cannot react that fast so the clock speed regulating muscle actions is a less degree of enhancement yet still faster than input. In humans it is 10 samples per second per pathway. Faster memory is sampled less to output to motion.

So how does a one half second sample of input turn into a fluid visual experience or a vibrating aural experience: distribution of starting points in time.

Centrally focused inputs, like the eyes and ears, the nose and the tongue are within close enough proximity to have a similar perspective.

Hearing inputs are focused through a single hole and single membrane that is pliable and transfers the wave of sound from one medium to another.

Visual inputs are focused through a single hole and single membrane that is transparent and transfers the wave of light from one medium to another.

Smell and taste are the same function but are less focused and directly stimulated. Pressure and temperature are focused individually which is why pain is specifically located.

Each input represents a pathway of processing. Pathways are able to share components (neurons) as their timing is specific and remains the same throughout the pathway. If they did not, the human brain would be larger than the human.

Neurons may contain quite a few connections, receiving different sense pathways in varying clock points, which is why targeting a specific area of a brain, as a specific function is absurd.

The largest amplitudes processed in neurons result in the most observable fMRI image, while literally a split second later, a completely different sense pathway is processed with much lower amplitude.

The miraculous appearance of hearing in a 'vision center', or smelling in a taste center is nothing more than amplitudes for that creature, in that frame of perspective are different.

Stop staring at the obvious and ask what is not obvious. That is curiosity. And intellect controls its degree of minimum while, awareness in humans, controls its degree of potential.

The degree of potential for reactive brains is repetition.

Humans know repetition all too well.

The military lives and breathes by repetition. Its training is repetition to reduce self-awareness, so as to permit nothing but long-term memory based, reactive responses.

Some people are more inclined to be willing to undergo the reduction of self-aware to minimum, and others are slightly willing for a cause of its own, such as patriotism, while still others are not willing to part with their 'self' for any reason.

Some others missed the draft by one day and have mixed feelings about having joined the police department instead, but were too long-term based at the time to know the difference.

Reactive brains accept input in distributed pathways processed not simultaneously, but rather orchestrated in a continuous live process.

Modern computer parallel processing is a joke as it is nothing more than 1:1 all over again, at the same time.

Reactive brains accept input and present its pathway's input to that pathway's memory. The memory is running faster than the input, so each input is combined with a degree of more memories.

That turns a ½ second sample into a slower time frame, a more clear, greater resolution, processed in the same time frame as input.

A human's 30:1 half-second long-term memory processing speed is comparable to a modern computer taking every single 1 or 0 and turning it into 30 1's or 0's and processing that much more data for each input at the same clocking speed it is running at now.

Should anyone try that with a modern computer you can wait your turn for the output.

The brain's output is not very fast but it is in the same time frame as the input.

That is how a dog can react so fast to a serious threat, so that once the human observer becomes consciously aware of it: the event is over.

Depending upon the repetition of training in response to that sort of reactionary speed, and the subsequent conscious awareness that results, will determine whether the potential for output is contemplative or reactive.

Intelligence in reactive brains, is based in the degree of enhancement, memory has over input. The subsequent output enhancement over input, determines the dexterity, clarity and degree of control output demonstrates.

Human Intelligence:

See above.

Long-term memory in humans IS reactive brain processing.

If left to run the show, the human will react just like the dog, or the cat or the reactive brain host of your choice.

The path of knowledge acquisition, experienced by the human is housed in long-term memory. It is not stored.

Computer memory is stored and retrieved and replaced. Brain memory is never in the same place twice.

Starting at the combination with input, in reactive brains, (your long-term process as well) memory is sent in pulses of amplitude from one neuron to the next in that pathway's route.

Neurons accept the chemical stimulus through the one-way requirement of a synapse, where it emerges as excitation of the lattice

182

of the neuron, to 'charge it', and is met by a pulse from another pathway stemming from the biological clock that combines the amplitude of the lattice with the amplitude of the clock pulse, resulting in an overload of amplitude for the lattice and the subsequent expulsion or 'firing'.

That process continues 'on down the line' as if it were a string with many knots and a drunk gnat skipping from one to the next and bouncing off the knot, waving in time with it.

Eventually, the gnat is so worn out that it doesn't 'charge' the wave of the string enough to be discharged and it is absorbed.

That would be the end of the line of a string of memory. The more knowledge the more use, the more thinking, the stronger the gnat is, the deeper it goes, the more neurons its needs to continue.

If we consider human, to be the Homo Sapiens species, where prior species of relatives were not Homo Sapiens, we would be right.

Homo Sapiens have only been around for a very few relative years (the oldest known skull of a similarly structured creature is approximately 150,000 years old) as evolution is obviously physical traits but not so obviously intellectual traits. A human is not what a human does. A human is what a human is.

And a human is aware of being human.

That is the non-zero degree of consciousness.

Consciousness comes from the second level of brain processing called the short-term process.

Short-term is just like long-term memory and works the same way as it is a degree faster than its input (long-term).

It too outputs to motion and is combined into larger pulses that should result in short-term control over output.

But that has to be learned.

Long-term output and short-term output are joined as the nervous system transmits amplitudes to muscle cells. The one with the highest amplitude is in control.

The progression of human evolution is more about second level, short-term brain processing speed, than anything else.

Not until the speed reached the degree to which it is in humans did the process exceed the space afforded to it. The skull grew to its fullest [internal] potential.

When that happened, outputs at the end of the memory pathways of short-term processing had no place to build another neuron, so they built another connection.

That connection started off being deep in the pathway.

Such depth was not capable of generating amplitudes of near equal intensity with the input coming from long-term memory.

As species of pre-humans evolved into the next generation of humans (and that process was continuous by procreation) the amount of knowledge increased and the intensity of new experience amplitudes joined with past memory amplitudes to push the connection higher and higher up the short-term string memory.

When it reached the depth of equality of amplitudes in short term, it become one.

That 'one' is you. It is the 'you', you perceive as yourself. A loop. The concept of 'now'.

It is consciousness. Being aware of existence.

In papers years ago, I attempted to explain the difference between reactive brain awareness and proactive brain awareness, as self and other awareness.

Self-awareness is actually a degree condition. It starts with the minimum of non-zero consciousness and increases to the degree of self-awareness the human makes happen.

It is the difference between testing an intelligence through reactionary long-term contemplation performing a Binet IQ test, or testing the ability to exceed long-term with pro-active short-term as in the IQ test at this site.

The speed of short-term in humans is typically 900:1 over input: a far advance for Homo Sapiens, and the tool for the ability to contemplate it, let alone explain it.

Reactive brain creatures are not conscious. No creature other than Homo Sapiens is conscious.
That does not mean reactive brains are not intelligent. They just, are not aware of it.

When your pet cat sleeps, he or she is experiencing memory without the benefit of input, from the current environment, which is their only relevance to 'now' and when awakening is abruptly thrown into this environment. Imagine waking up in a place you knew you were in before, but just not that moment.

When you sleep, the same process occurs, but you remember your sense of self, as the output to your long-term memory is your short-term memory.

Degree of intelligence is created through degree of advancement over input.

That degree emerges from the combination of return feedback loops in memory as well as the degree one level of brain is dominate over the other.

If children were taught how to identify and use their self-aware short-term processing, the rest of their lives would be spent contemplating, with little to no interference from past memory, as past memory would indeed be, known as the past.

And it would not rule their responses and the next generation of humans will abuse itself less and consider its fellow humans more.

If psychiatrists would treat patients with the tool of increasing self-awareness and stop the horrible and barbaric practice of regression, patients would heal and move on instead of heel and remember and relive.

Just like intelligence is a two part whole, your body is a two part whole. The brain and its actuators and sensors given mobility by the cellular structure it uses, are matched with the non-zero wave amplitude of the dynamic system to form a living creature.

You are aware of your existence, and you can be aware of yourself, to a degree where past existence has no relevance other than reference, and your future is where you are headed, without the chains of long-term holding you in the past.

There is more to life than what other lives have already given to impose on yours. The power to make it happen is within you, as it is you.

If you suffer from depression, worry, fear, anxiety or any forms of long-term process interruption in your life of 'now', do this:

Each time you are trying to stay focused on now and are interrupted with a memory of the past, regardless of what it is, STOP ... catch it... say OUTLOUD "That was then. This is now!" then regain the self-awareness you were interrupted from.

Others may look at you weirdly for blurting out something with no reference to them, but who cares? It is your brain. Perhaps they could read this material as well.

Marking your past memory as past and not relative to 'now' will reinforce the sense of self and with each use will drop long-term memories lower in amplitude.

The self you know you are will emerge. You can keep the good memories alive by remember their lessons and reflect on the bad memories only to remember a consequence. You will not be living a past, that you control.

You will reach it.

To reach the root cause of a long held problem, dissect it from the top, removing each layer, with each layer connected to previous causes, which are made up of previous causes, until the bottom reveals itself and it too will be **then**, and not **now**.

No one can hold anything against those who try to help with the knowledge they have been taught, based in conjecture and theory, and only upon the obvious.

It is high time the not so obvious is considered without the long-term memory's insistence of retaining the obvious, as it is similar, and this is not.

But we can advance beyond that and realize that it is our brain and we can be in control of it: and if we are, we need no one's help to make it work for us.

Chapter Twelve
Feeling, Dealing And Healing

Not everything about what the brain results in, has been described in this book, yet most everything about the brain's system of functioning, has been described.

If you have chosen to read the book in parts, and not in order of chapter, you will have missed some important concepts. Go back. Read it again: in order. Or read it first, before reading this chapter.

Feeling:

Feeling is considered being conscious of an inward impression, state of mind or physical condition. That definition, offered by Merriam-Webster, among others, has resulted in the arrogance of human-only feelings.

To be aware of an inward anything, is the non-zero state of consciousness.

It is only in humans.

To impose the lack of emotions on non-human creatures, is a combination of ignorance and arrogance, the need to feel better than others, to display dominance, without understanding what dominion means.

There is no such thing as 'emotional intelligence'. Emotions do exist as emotions because humans are aware of their being different than conditions. That does not make them independent from conditions.

In reality, emotions are the result of conditions AND previous conditions. Only the 'now' conditions: are known by the human.

That non-now condition, the output result of long-term memory, is with us all.

To creatures other than human, it is the only output they have. To humans, it is the first output, the seed of short-term processing and not involved in awareness at all.

No matter how much buzz the term EQ has garnered in business, as they seek to find ways to pigeonhole job applicants, there is no such thing.

Emotions are the output of long-term memory processing that happens without the awareness of the human. To creatures other than human, emotions are simply the output.

EQ became 'cool' when a book was published, claiming it to

be real. It is not.

Feelings, emotions, those 'things' we all think, react from and feel controlled by are the output of our past, that make no sense to the 'now' of self-awareness.

The total solution to emotions that are troublesome is to know what they are and realize they are not 'now'.

Discussing emotions with reason and logic tends to make some defensive. After all, they are their emotions. No they are not.

They are the result of everything put into long-term memory that controls the person.

That means, emotions are your past trying to compete with your 'now'.

If they win, and the default state of not engaging short-term processing is a win for your past and a loss for your future, they will control you. You will not like it. You will not know why. You will seek 'professional' help in controlling them and fall for the slightest seemingly reasonable hogwash program, self-help solution and fictional book to address them.

This book has not asked you to fall, believe or accept anything.

This book has presented logic, pure and simple. It has called upon your reasoning abilities to think it through, compare what you know to what you are reading and compare what you have read here to what you knew.

Think about it.

If there were such a thing as emotional intelligence, there would have to likewise be:

Motion Intelligence
Speech Intelligence
Art Intelligence
Sport Intelligence
Breathing Intelligence
And many others…

The output of your brain controls each of these functions, as well as emotions.

The only difference is that short-term processing is not

controlling emotions. They appear to be special and unknown: ripe for the picking by opportunists.

Where outputs of the brain are sent to motion in specific parts of the body, outputs are also sent to short-term where consciousness arises and self-awareness is possible.

One can cause the movement of a finger to type a letter or one can know typing so well that thinking the letter or even the word results in long-term output, where errors are only recognized after they are typed.

Emotions are the result of aural and visual long-term processing's output to short-term. You are aware of an emotion only because you are aware.

Creatures other than human also have emotions but they are not aware of them, any more than they are aware of anything else.

You have the ability to increase your awareness to the degree of knowing that you know.

Dealing:

Knowing that you know is the moment you realize that you are a part of a system, which is governed by a system that is duplicated in everything. By discussing the system, one is able to relate to others without the imposition of the past's hold on reality.

How one approaches knowledge, indicates the method they use to acquire knowledge. It is common sense.

A prime example was published as a paper of its own in order to find the response.

Common Sense:

Artificial Intelligence may as well be a branch of Psychology.

Both fields seek to understand the things brains do.

Where Psychology seeks to understand what causes desires and wants based on what desires and wants are observed, AI seeks to duplicate desires and wants based on what desires and wants are observed.

Desires and wants are called 'goals'.

Goals are the things observable that brains result in. Whether they are good or not so good (do not ever use the word 'bad' in the same sentence with 'good' if the desire is to not invoke ancient religious bigotries.) Good and bad are conditions of perception.

Psychology looks at goals as a good thing for order in life as the future must be the solution to the past?

Artificial Intelligence looks at goals as a good thing for a program to do to appear to be intelligent and builds a goal as an add-on module.

All of this is obvious deduction.

The not so obvious potential of the causes of goals is where both fields go astray.

Where Psychology looks into the past, causing old memories to be recalled, seeking the underlying single cause of a condition, Artificial Intelligence looks to the now, seeking to duplicate the single goal.

Add a good deal of those goals together and Artificial Intelligence believes it will result in a thinking machine.

Psychology can tell you that a large collection of goals means nothing if they are not connected.

At least Psychology has kept its focus by relying on theories of mind and brain, even though such theories are based in obvious observational deductions.

AI, on the other hand, has lost its focus.

Starting out as a science to attempt to replicate how the brain does things, AI has turned into a science that seeks to replicate what brains do, with no relation to how they are done, making all modules have no central connection.

In fact, mentioning how the brain works in Internet Usenet comp.ai, receives a refusal to post, as AI has nothing to do with how the brain works.

And that would be right.

AI has everything to do with how each AI researcher sees what the brain does, and how each AI researcher determines how best to duplicate that goal.

That makes AI the science of observational deduction and the process of making absurdity a common task.

If AI were AT (Artificial Transportation) it would look at what cars do and make motion machines, turning modules, stopping modules, speed modules: and if researchers in the AT field sought causes they would make galloping modules, running modules, and trick modules.

If AI were AP (Artificial Politics) it would make pork modules, deception modules and speech modules and if AP sought causes they would make power modules, conservative modules, liberal modules and anarchy modules.

If AI were AM (Artificial Music) it would make tone modules, timbre modules, volume modules and if AM sought causes they would make piano modules, guitar modules, drum modules, violin modules and top it off with a conductor module.

It is impossible to make a correct result without first making the cause, or at least admitting there is a cause.

The cause of anything 'intelligent' is a brain.

AI continues on, defining its science as whatever the currently in vogue method might be with no regard for the brain.

It is logical to observe and deduce. It is not logical to stop there.

What one sees is not what causes the thing being observed.

The result has been promise after promise, with not so much as a single truly thinking machine since AI was created.

Do not tell that to an AI researcher.

AI researchers are like sheep. They follow the leader. The leader is whoever has received the most agreeable assertion of fellow sheep.

Psychology finds an alternative thought and examines it for potential.

AI finds an alternative thought and attacks it for not being normal.

Anything that remotely appears to be an affront to the norm of AI is therefore wrong.

That is what happens when the science of replicating intelligence, fails to use it.

Intelligence is not relying on what has been done: it is evaluating what needs to be done.

AI has carried on with the illusions they create as if they real. The problem with AI: is that they are not.

In "Artificial Intelligence Group" in yahoo groups this happened:

"Any project that proposes to build a thinking android _IS_ an

192

AI project... (unless you plan to steal someone's brain! =P)

and:

"Well, let me clue you into something: By a typical estimate of the power of the human brain, you will require around 500 high-end machines and those will produce enough heat in a day to heat my home for a month in Febuary!!!" (Spelling, not corrected.)

and the best of them all, from sciforums.com in response to reading this piece:

"While it may be true in some respects, it seems like an awful lot of doublespeak to me. It's hard to articulate what exactly feels wrong about this site in general...maybe someone else agrees and can say it better. For one thing, they assert claim after claim, but they present no supporting evidence that I can see. This isn't to say there is none, but how am I supposed to weigh anything they say against my current beliefs if all I have to go on is their word that this is true? And if there really is no experimental support for their claims, then aren't they just basing them on their own 'obvious observational deductions'?"

"One of the things that bothers me the most is:"
quote:

The   cause   of   anything   'intelligent'   is   a   brain.

"Says who?"

NOTE: The link is no longer active as it was removed within two hours of the posting of this chapter as a review able paper: the entire thread, as well as every posting containing the link to this chapter. Not just this comment.

Artificial Intelligence is not a horrible mistake. It is a well-meaning mistake. It is a mistake at all, as it is based in guesses and subjective interpretations of observable things. It has no basis in fact only in vision.
In   "Why   A.I.   Is   Brain-Dead"   from   Wired

http://www.wired.com/wired/archive/11.08/view.html?pg=3

Marvin Minsky, co-founder of the MIT A.I. Lab, remarked, "There is no computer that has common sense. We're only getting the kinds of things that are capable of making an airline reservation. No computer can look around a room and tell you about it." As Wired stated: "the field of AI has lost its way".

"Researchers are making little progress developing computers with any knack for reasoning." [Wired]

In order to prove the point of one option for dealing with the knowledge of the brain, this piece was published at EnticyPress as "The Problem With A.I." before it became part of chapter 12. It was 'advertised' in online sources where AI researchers would frequent. The short advertisement for it included the line, "If you are an AI researcher, don't read it."

In dealing with knowing about the brain, the options are to be skeptical and learn, to be skeptical and refuse to learn or be agreeable to anything and ignore learning. It is 'common sense'.

'Common sense' is not knowledge.

It is the process of acquiring and using knowledge.

An agreeable definition of 'common sense' has eluded as each person believes it to be referring to knowledge. It is not. It is referring to the process of acquiring knowledge. A 'sense'.

The degree of any, is directly related to the degree of belief, based in the method acquired.

In schools, those who do not know, will find each other and use the replacement belief they have for knowledge, to find common agreement with others, and use the power of that association, to wield power over those who disagree, or do not fit the common agreement.

They are called, gangs.

Agreement for 'common sense' takes the form of one of these options:

A: **be skeptical and learn:** "The important thing is not to stop questioning. Curiosity has its own reason for existing. One cannot help but be in awe when he contemplates the mysteries of eternity, of life, of the marvelous structure of reality. It is enough if one tries merely to comprehend a little of this mystery every day. Never lose a holy curiosity." Albert Einstein (1879 - 1955)

B: be skeptical and refuse to learn: "Prejudice is the child of ignorance." Hazlitt

C: be agreeable to anything and ignore learning: "None has more frequent conversations with a disagreeable self than the man of pleasure; his enthusiasms are but few and transient; his appetites, like angry creditors, are continually making fruitless demands for what he is unable to pay; and the greater his former pleasures, the more strong his regret, the more impatient his expectations. A life of pleasure is, therefore, the most unpleasing life." James Goldsmith

Types A and C make up the majority of school children.

The type A common sense, requires contemplation to be skeptical and will result in learning and comprehension if the teacher satisfies the contemplation. These are the students who ask questions in class.

The type B common sense, requires rejection of learning, usually as a defense to the ignorance of not knowing or the defense of a belief. Ignorance will defend itself at all costs. These are the students who do not participate in class.

The type C common sense, requires nothing. These are the students who do what they are told as that is what it takes to get out of class.

"Its only common sense": spoken by most people in their lives, tells what form of common sense it is.

AI researchers are mindless sheep.

Two days after this piece appeared on the net as an article, the above 'thread' appeared at Ai-Forum.Org.

The advertisement was not posted at that site.

The title includes the term 'mindless': (not used in this piece). The poster, hiding behind the name 'Ribald' (which is obviously not a real name, as those in the dungeon rarely divulge their true identity), came away from reading this piece as its own paper with the impression that 'sheep' meant 'mindless'. (Type B).

It went on to read:

"I disagree because I see nothing more in his article than Lee's typical assertion that anyone that does not believe what he is selling is an idiot. After having seen his "theory of mind" it seems clear to me that he has no supporting evidence nor corroboration."

"His entire belief system (deliberate phrasing) is that the truth of his vision is self-evident, and that anyone that doesn't agree is either mentally incapable of understanding, or is deliberately refusing to see what he sees. I consider it far more likely that one individual is delusional than that the entire scientific community is."

"The process of peer review is fundamental to science, and is also practiced by Ph.D. level psychologists, contrary to what the article indicates."

The term 'peer' is not used in this piece. So how did 'Ribald' manage to connect 'sheep', 'mindless' and 'peer review', when only 'sheep' was read?

"The peer review process is what creates a firm basis for building the scientific knowledge base. The refusal to accept one individual's assertion without any evidence whatsoever is not an indication of foolishness as the article indicates, but is rather just the opposite. A theory that can offer no defense beyond attacking the critic is nothing more than an opinion with a large emotional investment by the author."

Excuse me? Type B has described his own posting. Peer review was never mentioned. Mindless was never mentioned. Therefore his deductions were never eluded to, yet he defends them.

Since he brought it up, the process of peer review is wonderful for the dissemination of accurate knowledge.

It is horrible for the dissemination of new knowledge as the peer review process is Type B all the way. Its purpose was to assure accuracy while its result is to assure redundancy. Peer review results in mediocre science taking center stage through agreement of the sheep.

As this book has shown, countless times, perception of visually thinking people requires visual stimulus.

Visual stimulus only relates to what has already been visually stimulating. If it is not equal to the known image it is wrong. It is the primary reason science makes progress so slowly, innovations are the norm where invention is the anti-norm. Peers are therefore how science defines sheep?

Quite a few persons linked in from the Ai-Forum.org board to this piece as its own paper, as every access has been tracked upon log in, the link is evident.

'Ribald' continued:

"I think that truth wins out in the end. Infamous is only equivalent with famous if you are a rock star or an actor, it doesn't work for scientists unless your work has merit."

"If you get attention for your ideas only to have them dismissed as nonsense you have gained nothing."

"The 'stir it up in the hopes of making me famous' approach always backfires in the end. Unless of course being a sad footnote is sufficient to stoke your ego, then it might be effective."

"Merit" means "individual significance or justification" [m-w] which is a perceptive quality.

'Ribalds' perception appears to be 'get attention' based in 'stir it up in the hopes of making me famous'. Perception has a way of showing motive of the perceiver.

After all why would a thread be started in a forum where no reference was made to the topic; be started based on a misconception of what the article said, deduce incorrectly what it meant, yet retain the expected form if it were not for the person posting the delusion to gain 'peer' acceptance or to find common agreement with others, and use the power of that association, to wield power over those who disagree.

'Raphael', who appears to be the site owner or moderator at Ai-Forum.org then stated:

"I must say Ribald would not have begun this thread to create an opportunity just to say mean things about someone; he knows that kind of thing would only be deleted here. So he must have seen some merit for discussion regarding the linked article."

"Therefore, let the *article* be discussed...AI researchers are mindless sheep, pro or con..."
(That would be nice if it had anything to do with the article.)

"Ribald says he disagrees with the article because he doesn't like the guy who wrote it since the guy doesn't play by the established rules of the rest of the community. Pennywise implies his disagreement because the author and others don't play by the established rules of the rest of the community."

"Is anyone missing the irony displayed thus far?"

Apparently so.

'Ribald' responded:

"I started the thread because I think it doesn't hurt to examine criticism. I also think that too many promises and claims HAVE been made, which is one of the few areas in which I agreed with the article."

(Backing off slightly to mend fences?)

"I will assume that your assertion of my motives should be taken at face value instead of as an implication."

As it is obvious the thread, appearing out of nowhere, was indeed intended "to create an opportunity just to say mean things about someone".

"I think that your characterization of what I said is perhaps colored by your own feelings about the AI field Raphael."

"What I said was that:"

"1 - I saw the article as actually not being about AI, but instead being about justifying a lack of support from the community for a theory that has no evidence."

And words that were not in it. Not to mention the point being missed and therefore proven by its absence.

"2 - Psychologists go through the same peer review process as other sciences, contrary to the assertion in the article."

It never said peer anything and referred to the intellect of psychology, not the process.

"3 - Peer review is a necessary part of science, and does not indicate "sheepism". It in fact represents an active analysis of new theory. Failing to do so and just believing anyone's claims sounds much more like a sheepish behavior to me."

"Sheepism". Excuses for refusal to learn. A type B common sense cannot learn that would end type B.

"I take it you see it differently?"

'Raphael' responded:

"Actually, I'm with you all the way. My earlier post was intended only to help this thread address the topic - the article - not its author (or any other names that might be pulled in). I'm a big proponent of targeting issues in discussion, not people."

This comment sets the site owner apart from the fray yet in agreement, so a leadership role is assumed.

"Upon re-reading this article, however, I do see that in this case, the author has included himself so tightly that there cannot be any purely objective division between the two.

In fact, it would appear the only motivation behind the writing was the rejection of a post to a newsgroup and some undesired responses to other posts made by the author in other locations. So he used his own website to vent his frustration. *shrug*"

It will be noted that the point of the article has not sunk in. How could it?

"Your last post, Ribald, clarifies your points excellently, and thank you for that."

The thread ended (as of 30, July 2003) with 'Halion's desire for inclusion and proof of worthiness.

"The man's paper is a glaring fallacy. Generalization."

"Since I personally try to figure out why a brain does what it does, I am a fact, contrary to this mans argument."

"Let this man's paper lay with the rest of the mindless writings of a frustraighted hothead."

So we have agreement amongst the gang, caused by not reading the concepts presented, only the words: and resting at the point of agreement in hatred and disdain for anything not already known, followed by the second exclusion from the simple member status to the 'special' member status. Not one of them, understood the article.

The more any person believes in something the more they defend it. The more they defend it the less they think.
"Against logic there is no armor like ignorance." Laurence J. Peter (1919 - 1988)
"Nothing in all the world is more dangerous than sincere ignorance and conscientious stupidity." Martin Luther King Jr.
"It is impossible to make people understand their ignorance; for it requires knowledge to perceive it and therefore he that can perceive it hath it not." Jeremy Taylor (1613 - 1667)
"Discourtesy does not spring merely from one bad quality, but from several--from foolish vanity, from ignorance of what is due to others, from indolence, from stupidity, from distraction of thought, from contempt of others, from jealousy." Jean de la Bruyere (1645 - 1696)
"We want the facts to fit the preconceptions. When they don't, it

is easier to ignore the facts than to change the preconceptions."
Jassamyn West

AI only sees what it can see.

It sees the obvious and stops there.

Intelligence, among other things, is the ability to at least look for the not so obvious.

How do you deal with knowledge?

Do you excuse it away so as not to offend your past?

Or do you question it, so as to clarify your future?

Or do you ignore it, so as to retain the comfort of your past?

Healing:

"Make your own recovery the first priority in your life." Robin Norwood

Depression, fixation, and reaction: and most other maladies of the brain's dependence on long-term processing (your past) need not be your future.

"The trouble with our times is that the future is not what it used to be." Paul Valery (1871 - 1945)

By allowing your past to rule your 'now' you are dooming your future to be just like it.

The path to recovery starts with knowledge.

The knowledge of how you think. The knowledge of what thinking is. The knowledge of how you are aware of thinking. The knowledge of how others think. The knowledge of your mind, prepares you to use it.

As an adult you have responsibility to mold the future in your children.

Treat your child, like the investment your car is.

If your car does not start, check under the hood, see if it has fuel, check the battery: don't beat it.

If your car will not move fast enough, give it more fuel, clean its carburetors, change its tires: don't yell at it.

If your car is dirty, wash it: don't scream at it for getting dirty.

If your car is getting older and not running on its own, tune it: don't ridicule it.

If your car is thirsty, give it fuel: don't make it get fuel for you.

If your car has a hard time getting up a tough hill, push it: don't belittle it for not being strong enough.

If your car keeps getting lost, check the driver. You are your car's parent even though you are not your car's father or mother.

If your car keeps bumping into things or causing 'accidents' get the driver some help.

If your car goes too far remember who the driver is and employ a little restraint before the time to stop, instead of realizing afterwards, the car only did what you showed it to do.

But by all means, never hit your car. It will learn that hitting is the reaction to all things misunderstood and you will face the result of your driving's future.

Dents in your car can be pulled out and painted over. Dents in your child may be painted over but they will remain forever.

Teach your child how to think, so your child does not need your thinking to live by.

Teach your child to be self-aware so your child knows they are not your past.

When a car gives up, we put it in the junkyard, grind it up and make more cars. When a child gives up, the species loses hope just as much as the child does.

The Brain **IS** A Wonderful Thing.

It controls everything you do.

If you control it, the brain becomes a tool and you become whatever you desire. If it controls you, you become the tool of your past and whatever the past desired.

It has been the default condition of humans, not understanding how the brain works, to fall victim to its process and live the past all over again, with every new experience.

You are your brain and its mind. You are not your body, your color, your race, your creed, your preference or your illusions or ignorance.

It is the goal of The Enticy Institute, to use the knowledge of the brain, to help humans live better lives and to work and live together as humans and not long-term only reactionary creatures.

It is a goal that has waited long enough.

Chapter Thirteen
The Biological Clock
And Time

In the 18th-century Swedish botanist Carolus Linnaeus grew a garden that told time.

In that garden flowers opened or closed their blossoms an hour apart around the clock . Each one carefully chosen to perform the function it was controlled by.

That function has been known for centuries.

Science has chosen to call it, circadian rhythm.

So what makes the rhythm of the plants differ from us?

Or does it?

Gonyaulax polyedra is the name given to colonies of the microscopic alga luminescent at night, especially if the water is agitated, and relatively nonluminescent during the day that illuminate ocean waves at certain times of the year.

Under constantly dark laboratory conditions, this regularity of rhythm of luminescence and nonluminescence continues.

Our planet can be said to go through certain regular patterns of rhythm as well. The rising and setting of the sun has been a noticeable event since man first began to comprehend duration of events and then the necessity of regularity in planning and farming.

Thus the concept of measurable time was discovered and implemented using first, the moon and then the Sun's regular pattern. At that time the Sun was indeed considered to be rotating around the earth.

When applied to humans though, the regularity of rhythm is neatly tucked away in the examination and study of sleep patterns and the belief of the reality of time.

It had been the rhythmic activity of most living things at certain times of the day and night that offered strong support for the existence of biological clocks.

When scientists first studied these rhythms, they dealt mainly with the rhythmic leaf movements of plants.

Many plants go through a rhythmic daily cycle--their leaves are extended during the day and droop or are folded at night.

When such plants were kept under laboratory conditions of constant darkness or low intensity light, the sleep movements continued for days.

Under these conditions, however, the frequency, or time

required to complete a cycle, of the rhythm was not exactly 24 hours.

Afterward it was learned that the daily rhythms of many living things continued when they were subjected to similar laboratory conditions. For most organisms, the frequency of the persistent rhythm varied from 23 to 27 hours.

Since these rhythms had a frequency not exactly 24 hours long, they were called circadian rhythms, from the Latin words 'circa', meaning "about," and 'dies', meaning "daily." Or about 24 hours. Almost daily.

A pattern of study had already begun to emerge.

What was being studied as internal clock frequencies was being imposed to have regulation by the Sun. 24 hours.

And when it was found that nature does not adhere to what man had invented, a name indicative of that was not chosen. One that was indicative of not meeting man's expectation was chosen.

Since the scientists were determined to find an adherence to their concept of time in natural events the name seemed appropriate.

But why 24 hours?

That is one whole day as measured on earth.

Why not the parts of that day?

Why did the scientists choose to observe in relation to a full rotation of the planet?

Because the events they observed seemed to follow in conjunction with day and night. Flower pedals opening in day and closing at night.

If those same flower pedals had opened in the morning and closed at mid day and opened in the evening (which some do) would the name have been based upon that sequence of events?

What if the scientists bothered to check that day and night are not equal 12 hour parts?

What starts out to attempt to describe an event or a situation study, becomes the definition.

Circadian is the prime example.

Who says clock rhythms are based on a day?

Definitely not the plants.

Who says any name is clearly responsible for defining an entire theory?

Who says relativity does not define the observer and not the observed?

Sorry to say, that it is the use of a name that lends to it's credibility, and then the study itself is no longer the primary issue.

Circadian rhythms at or near 24 hours studied in plants led to the study in animals.

In experiments on rats, Matthew A. Wilson of the Massachusetts Institute of Technology and Bruce L. McNaughton of the University of Arizona inserted electrodes into the hippocampus, a region of the brain thought to be involved in spatial memory. As the rats learned to navigate a maze, Wilson and McNaughton concluded that their neurons fired in certain patterns corresponding to specific parts of the maze.

But did they?

Firings corresponding with any singular event would indicate a similar event but would a firing indicate the event was the visual notice of a corner or would it indicate the touch receptor's acceptance of the texture of a certain part of the maze's floor? Or does the brain take complete pictures including sound and smell and touch of an event and store and process it all in the same place like a computer?

For several nights after the rats' maze exercises, their hippocampal neurons displayed similar firing patterns; the rats were apparently playing back their memories of running the maze. Or were they? Were the rats playing back visual reception of a corner that matched the same of another corner that matched the same as the corner of their cage or were they dreaming of the end of a straight line?

Wilson and McNaughton noticed that the major difference was the firing was more rapid, as if the memories were being run on fast-forward. The firing occurred during slow-wave sleep, a phase of deep (but not dreamless) sleep marked by low-frequency pulses of electrical activity in the brain.

What did Wilson and McNaughton observe? We'll see.

Then the studies of humans:

A team led by Avi Karni and Dov Sagi at the Weizmann Institute of Science in Israel trained volunteers to quickly recognize the orientation of symbols hidden in images flashed at the periphery of their vision. The workers had previously noted improvements in performance over a 10-hour period following a training session.

To determine whether sleep played a role in this phenomenon, Karni and Sagi disrupted the sleep of volunteers after they had had their training session. Interfering with the subjects' slow-wave sleep had no significant effect. But an equivalent disruption of REM sleep, which is marked by rapid eye movements (hence its name) and vivid dreaming, kept the subjects from improving overnight.

What was it that Karni and Sagi observed? Was it something totally foreign to the foundations made by Wilson and McNaughton?

In an article appearing in Scientific American October 1994 Volume 271 Number 4 by John Horgan; Karni and Sagi were quoted as saying, "These results indicate that a process of human memory consolidation, active during sleep, is strongly dependent on REM sleep,"

The article then concluded that the experiments lend support to a theory advanced by Jonathan Winson of the Rockefeller University that dreams represent, in effect, "practice sessions" in which animals hone survival skills. Why did Karni and Sagi detect memory consolidation during REM sleep and Wilson and McNaughton only during slow-wave sleep? The answer seems to be that each group studied a different type of memory, one involving a highly repetitious task and the other the recollection of a place?

Neither research group considered the one single monumental hurdle to understanding the brain. Instead of considering that one observation was the result of one level of processing and the other result was the observance of another level of processing both groups assigned a determination based upon what they assumed.

The repetition of a task and the recollection of a place.

From a study of clock rhythms, the circadian process evolved into a study of sleep patterns which would be comparatively judged by circadian rhythm theory. So the process that began by examining regularity in nature was converted to study something totally different. And that is where circadian rhythms remain today.

Time Piece vs. Regularity

In science, the measurement of time involves two steps: specifying the exact moment when something happens, and establishing a standard interval of time, or how long something lasts. Various devices--including sundials, watches, and clocks--have been developed to indicate time and to measure time intervals.

It is when something other than time is at work that science becomes confused.

Yes.

Confused.

Examine the turtle.

It can represent a device.

Our turtle leaves the turtle farm running (sorry, make that scooting) for it's very life at exactly 3PM Central Standard Time.

It's destination is unknown.

It will stop when it no longer senses fear.

How long will it take to no longer sense fear?

The notion of time can not answer a question when the parameters are not present.

The question of how long something lasts cannot be determined. Fear will subside as part of memory processing which is the cause of the notion of time.

So it is with examination of the events that are a part of the process of thinking.

There are other mitigating factors involved in thinking. Those other factors are more observable than what may or may not control them.

It has been easy for science to declare that if something is not known and no one has determined how to make it known then it is random. Without cause. That is the ineffable.

In conversations on a marvel of modern technology, the internet newsgroup I have had the pleasure of introducing thought experiments to people in other parts of the world. Such contact is priceless as far as I am concerned and it has resulted in quite a few interesting 'threads' as they are called. Conversations in open and closed (email) cyberspace.

One such conversation was regarding my assertion that the brain is controlled by a single clock frequency. I had forwarded a copy of a Boston Globe article: FRIDAY October 7, 1988 PAGE: 8 By Judy Foreman, Globe Staff entitled BOSTON TEAM PINPOINTS MAN'S *'BIOLOGICAL* CLOCK'*. Which read...

"A team of Boston researchers has pinpointed the mechanisms in the brain that underlie the ticking of the human "biological clock."

The "biological clock", located in an area of the brain behind the eyes known as the SCN is believed to govern a number of important daily biological patterns, known as circadian rhythms, including the sleep-wake cycle.

Scientists believe that this clock is reset every day by "entrainment" or synchronization with nature's own cycle of lightness and darkness.

When circadian rhythms become disturbed, such as by crossing a number of time zones in an airplane or working at night and sleeping by day, temporary derangement of the biological clock is at fault, according to circadian theory.

The research, published in the journal Science, was led by Massachusetts General Hospital researcher Dr. Steven M. Reppert, also associate professor of pediatrics and neural sciences at Harvard Medical School and Edward Stopa, Tufts University School of Medicine associate professor of neuropathology and also a neuropathologist at McLean Hospital Brain Bank.

The MGH-Tufts team found that the SCN, or suprachiasmatic nuclei, contains specialized receptors for capturing a hormone called melatonin, which is made elsewhere in the brain in the pea-sized pineal gland.

Curiously, they said, these specialized melatonin receptors seem to exist only in the SCN and nowhere else in the surrounding brain region called the hypothalamus, which is thought to regulate pituitary function and play a major role in triggering the flight or fight

210

response, rage, fear and other emotional reactions.

The new finding, Reppert and Stopa said in interviews, buttresses other recent findings showing that giving melatonin to people suffering from jet lag seems to alleviate their symptoms. Now that its biological site of action is better understood, they added, melatonin should also be useful for treating sleep problems of shift-workers and blind people and possibly helping newborn babies get onto an adult sleep-wake cycle.

The MGH-Tufts team also found that melatonin receptors exist not only in the SCN of adults but in the brains of fetuses as well. They studied fetal tissues obtained after therapeutic abortions. Although fetuses, unlike adults, cannot respond directly to light-dark signals, melatonin from the mother may provide the fetus with important time-of-day information, they added.

To which a response was eagerly anticipated. I will refrain from identifying the correspondent out of a kind gesture to his stature and healthy professional respect for the author of the following.

Subj: biological clocks Date: 95-11-16 09:38:41 EST

"I'm familiar with the concept of a "biological clock."

In fact, more than one has been discovered. This is not at all the same thing as applying the concept of "clock speed" to a system of neurological activity (like the brain). The former implies a periodic glandular activity which in turn stimulates changes in neural activity in certain parts of the brain.

The latter (as outlined in the excerpts you provided) involves neural activity whose elements are coordinated with some common frequency.

This is not the case. Neurons do not operate according to a common clock signal; they respond individually to the stimuli presented to them, and thus asynchronously. Impulses speed up or slow down in this asynchronous manner depending upon the strength of the stimulus. Graded potentials, on the other hand, do not result in changes in firing rate but in quantity of neurochemicals released.

The result in both types of transmission is a change in the amount of neurochemicals released per unit of time by individual neurons, either because there are more (or fewer) releases of a fixed amount of neurochemical per unit time, i.e., firing speeds up or slows down (impulse); or because more (or less) neurochemical is released per fixed firing rate per unit of time (graded potential).

Even in the latter case, this fixed firing rate is characteristic of particular cell types, with some considerable variation among individual cells within a given type. Neurons are quite anatomically variable, even sometimes within a given cell type (e.g., number of appendages and connections) and this has a physiological affect.

The important thing here is that all of these processes are more or less independent as far as individual neurons are concerned. The concept of clock speed would only make sense if human neurophysiology operated along the lines of a synchronous digital system, which it doesn't.

The mere fact that periodic events occur (such as the release of hormones which affect sleep or wakefulness) does not mean that the neurons so affected are synchronized -- it simply means that they are activated. General activity, though occurring at the same time, does not mean that the individual elements are synchronized.

Imagine the difference between a group of soldiers who are given orders to march in step at some given speed (perhaps by the "clock signal" of an officer counting-off), and the reaction of a crowd when some of its members are sprayed with buckshot. There will be simultaneous activity in both cases, but only to the former (formal marching) could one apply the concept of clock speed."

As can be seen from this 'thread' quotation the author is living in a digital world whereby things are compared to digital and knowing that the brain is not digital finds any reference to anything possibly also used in digital to be erroneous.

Furthermore before disclosing my response it must be evident that the author is also suffering from the dreaded observational illusion bug. Seeing the process as the outcome and not the process as 'A' process, part of a 'system'.

In my response:

"And where pray tell does this stimuli come from?

From whence is it submitted?

Is it submitted by chance?

Is it submitted by ruse?

212

Is it not submitted?

It just happens?

Does something responding to an order to respond make it less than something that acts upon something when it receives it?

Does it make it different?

The strength of the stimulus can be seen as the value of it. In a chemical process the strengths of the stimulus would indicate the value.

So a value of (0) would appear to indicate non firing. Numerous stimuli firing in an order of (0) and other variable values would appear to be value based when it is actually system based. A whole bunch of values would be observed as the neuron fires upon the presence of the strengths that does not preclude the value of (0) which does not cause a neuron to fire as there is nothing to send further. Thus keeping the firing pathway in (0).

The observer 'sees' what appears to be non activity when it is in fact low values. Activity continues unless of course the subject is DEAD.

Individual neurons are parts of the whole.

If they were more or less independent then there would be no binding and if there was no binding not even you would be you you'd be all of you'all. The whole is it's parts but the parts are not the whole they just make it up.

The brain of living creatures is not digital as you say. It operates as a synchronous analog system. It's called a multiple serial pathway synchronous parallel system.

In reference to: "The mere fact that periodic events occur (such as the release of hormones which affect sleep or wakefulness) does not mean that the neurons so affected are synchronized -- it simply means that they are activated. "

Would mean that there is no logic to the system that determines logic.

Which means that chaos determines logic: which is the opposite of the truth. It takes logic to determine if chaos is present and it is the lack of input that precludes logic from inferring logic

213

upon seemingly chaotic behavior.

The living brain has far too many inputs to ignore efficiency.

A neuron in a living brain is used by more than one pathway of processing.

The synchronicity keeps things in order.

The appendages and connections are indicative of the multiple use of a particular neuron not of it's being any different.

Drugs and other foreign substances and occurrences interfere with that order and in doing so disrupt normal thought patterns.

A brain born with a deficiency in pattern order will make disassociated connections and result in observable deficiencies in output. A planned drug therapy will interfere with that order and in doing so potentially correct a deficiency.

The concept of clock speed is also applied to the reaction difference of the crowd.

The one nearest the blast receives the impact sooner.

The one farthest receives the impact later.

The one closer the muzzle receives the audible report sooner the one farthest receives the audible report later, so upon observation the group appears to move as a unit in response but in fact the result is a wave effect.

The officer counting off the clock speed is no different than the muzzle explosion.

They are both causes. Singular causes.

The reaction of the soldier's to the officer is one based on known and reinforced action to a known and expected cause.

The one associated with the blast is a startle and not a known cause, making each individual group member react according to his previously inputted reinforced reaction to startling events with the potential outcome of personal harm. No two persons will make the same reaction. Once again the observer sees a group reaction in both cases but in fact the reaction is singular on all cases orchestrated by a singular causative action, distributed.

So therefore a disinterested observer would declare the group to function by it's outcome where an interested observer would declare the group to function by it's motivation.

Every input receptor of the brain acts as a singular motivator for further processing. Each input's variable values which cause the calculative chain of neurons for that receptor to activate the process is regulated in it's firing order by a synchronicity to a single

214

motivating clock frequency. Each step along that pathway is another syncratic breakdown of the same base frequency.

Let us examine a simple experiment if you will: Your wrist watch. (let us assume it is a quartz movement.) It functions at 60 clicks per second for the second hand and 60 clicks per hour for the minute hand and 12 hours per clock face rotation of the hour hand. All of which happens twice a day to make up 24 hours of telling time. What is the frequency the clock operates at? It would not be 60 as it would not denote a clock face rotation and that would account for only seconds. It would not be 60x60 as that would account for only seconds and hours. It would be 60x60x12 or a frequency of 43,200.

The second hand advances once every 43,200 waves and we see the seconds change slowly as it ticks along at increments of one second. The minute hand advances once every count of 60 sets of seconds or every 2,592,000 waves and the hour hand advances once every count of 60 sets of minutes or 155,520,000 waves. So 43,200 is wonderful. For a clock.

If you were an uninterested observer you would look at the clock face and determine that there were three hands operating in what appeared to be a regular basis but using the soldier and crowd analogy there could be no singular causative action for all three since they function in different rotation speeds.

Put the idea of that wrist watch into turning it into a hypothetical table clock that made an audible click with ever second, minute and hour. Now gather up hundreds of thousands of clocks the same make, model and sound and set them side by side and on top of each other in a giant room. Connect them all to ONE quartz oscillating frequency. But stagger them by one wavelength of the base rate so they won't all click at the same time. Each wave of the frequency would be assigned to one clock, some would be assigned to the same wave of the frequency as there would be more clocks than frequency wavelengths. Then stand aside and cover your ears as the clicks would seemingly cease to make sense as they all blended together into what would appear to an uninterested observer as a hodge podge of noise when in effect it is a synchronous symphony of sound. We might call it pink or white noise.

To the interested observer it is indeed a symphony. Something else would emerge as well. Every clock tick that matched another clock tick would cause that simultaneous click to set up it's own wave function. Clicks would be heard to oscillate at a regular louder and measurable rate. Some of those waves would be measured and someone would build a machine to do that and would call it an EEG. The result of joint frequency firing order would be measured while the actual frequency calling the shots and the frequencies resulting in the clicks would be too quiet to be noticed and housed too deep within the room of clocks to be easily measurable by anything other than room surgery. Until Francis Crick identifies the click that is a harmonic of the analogy's hour hand movements.

Replace the central quartz controlling movement with one who's frequency functions in reverse to the clock analogy. Replace those clocks with receptors and have each receptor send an inputted value once every 17,500 waves.(Two times per second.) Replace the clicks of hand movements with neurons at subconscious level and fire them at once every 583 waves and in the conscious level of short-term processing, once every 19 waves and you have an understandable representation of the syncronicity of the living brain's biological clock.

The brain's clock breaks down in ever faster computation whereby the clock extends in ever slower computation. It's upside down. But the brain must do more than click. It has to process. So each click represents the firing of a neuron while the value the neuron processes and the synapse transmits varies by a process that is described in a simple formula. The process itself is electrochemical and charge state processing, not electronic.

The uninterested observer then sees the above illustrations and declares them not to be indicative of brain functions. "

And his response:

"On the contrary, it is quite different. The concept of clock speed cannot be applied to the asynchronous units. Talking about causes simply confounds the issue."

Excuse me here, but if there is no cause then there is only chaos and if there is only chaos there is only random order and that has been more or less excused from scientific thought for quite a few days now. The brain does not 'fire' at random. But there is more:

"You evidently fancy yourself to be some sort of master of psychological manipulation. It is not a delusion shared by others. "

There was more but it got personal as it always seems to digress to when the correspondent fails to comprehend the argument and can not think of an intellectual response. I do not feel a desire to embarrass the author.

An article I was unable to forward to the correspondent who digressed to apparent feelings of insecurity was the following from The Boston Globe. FRIDAY, April 29, 1994 PAGE:3. As provided by the Associated Press.

ADVANCE REPORTED ON INTERNAL CLOCK GENE

WASHINGTON -- A gene for the "internal*clock" that sends the body wake-up alarms in the morning and that brings on slumber at night has been located in laboratory mice, a finding that may prompt a similar discovery in humans.

Joseph Takahashi of Northwestern University, senior author of a report to be published today in the journal Science, said the research could lead to drugs that will overcome jet lag, keep night workers from falling asleep on the job and solve narcolepsy, one of the most common sleep disorders.

The biological clock, located in the brain, controls the daily, or circadian, rhythms of life. It triggers changes that invigorate or slow the body. It is the circadian rhythm that is disrupted by flight across time zones, causing jet lag. Circadian rhythms have fascinated scientists, and the work by Takahashi and his group is the first to locate in a mammal the gene that plays a key role in the cycle.

Takahashi said researchers in his lab located the gene by finding and then breeding mice that lacked the gene. The gene was located by a system that measured the circadian rhythm of 300 mice automatically at the same time. Takahashi said that exercise wheels in each of the mouse cages were connected to a computer. When each mouse awoke and started exercising, a switch was thrown that recorded the time.

"They all started within a minute or two of the same time each day," he said. Except for one mouse. Researchers discovered that this rodent started an hour later. Some descendants also started late.

By comparing the genetic pattern of the prompt and the tardy mice, Takahashi said they located a mutation in a chromosome. "

Imagine a room full of clocks. Now replace them with mice. Then notice that mice were the first experiments in circadian rhythms which led to such in humans which led to finding the neurons that

217

perform the function which led to finding a gene that controls it. Then imagine that perhaps along with the establishment of a biological clock and the potential of a frequency that "controls the daily, or circadian, rhythms of life"... Perhaps, just perhaps science has also been confused about what that rhythm actually is.

Time has been viewed differently in different cultures and at different times. In some religions, time--particularly human time on Earth--is thought to run in cycles in which people die and are reborn again and again. Some ancient Greek philosophers believed that time is an illusion and reality is unchanging and motionless. Some major religions teach that time was created and is destined to end one day in a terrifying climax. Isaac Newton believed in a time and space that were absolute, ideal, and unchanging. His vision was overthrown by Albert Einstein's theories of relativity, which required people to think of a space-time combination that contradicted common sense. Under no circumstances will this work manage to discuss either General or Special Relativity.

In an article appearing in the New York Times March 14, 1995 'Modern Life Suppresses an Ancient Body Rhythm' Natalie Angier writes"

"As the vernal equinox advances, and the sun lingers in the sky a bit longer each day and the buds poke forth like babies' fists from every barren twig, even urbanites may feel the pagan craving to revel in seasonal rhythms.

After all, the lengthening of the day and the warming of the air exert a tremendous influence on virtually every other life form, inspiring migrations, ending hibernations, inciting growth and exciting lust. Surely humans, too, must be prey to the power of the seasons, the return of light and the chastening of night. Surely people's innate circadian clocks must react to the return of spring, resetting themselves to keep pace with the extra daytime hours.

There it is again. The name making itself the definition. The above conclusion indicates the author, like most scientists, believe the internal or biological clock is controlled by light. By the presence of light or the absence of light. After all isn't light the day and lack of light the night and do not both together make up what should be a day and doesn't a day come close to the definition of the rhythm ? The almost daily rhythm?

There is much study to support the notion that light controls the biological clock. Referred to as the internal clock, - the part of the brain that responds to light and dark. But is it light that controls the clock or is it light that feeds the clock? And wouldn't the feeding of the clock tend to slow it down when there is less food and speed it up when there is more food?

What happens to all those completely unable to function eskimos during 6 months of 'night'?

It is common knowledge that when a light bulb is connected to a battery with a potentiometer in series between the two, the potentiometer will control the amount of power applied to the light and therefore make it bright or dim.

But just as the light responds to the increase or decrease in power through the control it also responds to, the battery's power supply. The less power in the battery the dimmer the light will get, and no amount of increasing the potentiometer will increase the light from a dying battery.

Likewise the battery can be recharged by applying current to the battery or to the line attached to the potentiometer.

The result is almost the same with one twist.

Recharging a battery by connecting it to another battery will drain the good battery.

But recharging the battery by connecting it to a power supply that is separated from the battery (such as in a recharging unit) will recharge the battery without draining the recharging source.

But we're discussing the brain here.

The brain contains memory of the rhythm set up in cycles of activity that correlate to actions and events of a normal and relatively regular schedule. When that schedule is broken, the brain continues to function as if it were still following the same pattern. What has come to be known by the obvious culprit in this phenomena as Jet lag occurs when the brain continues but input reality changes.

A good deal of study and many recent commercial ventures have been dedicated to applying the use of bright light to alleviate the stress caused to the brain by that change in reality. Patients are subjected to long periods of bright and even light that has been noticed to 'reset' the biological clock. When in effect the clock hasn't changed the brain, the brain has been retrained to accept the difference in reality by a steady flow of continuous equal values observed by the brain's levels of processing as another pattern and one that eventually equals the previous pattern. If it works. Each patient is different.

Could the results that jet lag study receive actually be the result of feeding the clock to the point where it catches up with the brain's perceived rhythm? Thereby 'resetting' that which is timed , the memory, instead of the 'timer' so to speak?

Light does indeed feed the visual sensors of the body and in doing so aids in the body's natural recharge that works in a faster and more stressful manner the same as the at rest period we experience with sleep (when sensors are not working to full potential.)

To understand why that is it is necessary to comprehend a very important difference between life and artificial.

Life lives. Artificial does not. That may seem self evident but the difference has been lost in understanding.

Life is alive and as such can not possibly permit an assumption of being 'off' unless that 'off' were something altogether different than we perceive it to be in our creation of an artificial world.

A clock is artificial.

It performs a function we humans designed for it to do based upon the way we perceive existence to behave.

It counts in increments upwards in longer periods of duration. Seconds to minutes to hours to days. But there are shorter periods than seconds.

And it takes quite a deal of precision to artificially create the measurement of shorter and shorter periods. It is the requirement of an artificial system. It can not be real so it has to adapt.

The living biological clock is natural and counts in increments downward in shorter periods of duration.

This guarantees a natural limit on one end of processing. That limit keeps the biological clock and the living creature it controls being what it is. Acting how it does, performing tasks and making decisions how it does.

Unlike the artificial clock the biological clock starts at it's maximum and divides. It is an established operating frequency. Increments of it are used to control differing levels of brain computations. Input receptors do their reception twice a second. Muscles do their movement functions 10 times a second. Between the 2 in and 10 out sequence is a host of varying levels and different clock speeds all derived from one maximum operating frequency.

So in effect it can be said, based upon one perspective, that the function processes of life are upside down to the functioning processes of the artificial systems we have designed to mimic life. From the other perspective it can be said that artificial systems are upside down to the natural way of brain processing and that is the stumbling block.

It is this upside down perception that has kept the greatest minds in science from 'seeing' fact. It does not appear when the vision is trained to see upside down.

It is well known that the visual process of the brain 'sees' what the eye's receptors send to it. And that 'sight' is upside down to the reality of the vision. The brain functions in opposite sides to it's inputs and outputs.

How then does the brain 'see' upside down and act right side up?

If you have ever sat upon a see-saw you know how.

Down goes one side up goes the other.

It is a transfer of energy applied to one side to an expulsion of energy at the other side. Upside down input can result in right side up output but it is the center post, the bridge of the teeter-totter, that is required to transfer the energy.

In the brain that center post is the process. But as unlike the evenly balanced teeter-totter where a 50 pound kid will launch his little sister, balance another 50 pound kid and hardly budge his parent the brain has 2 per second in and 10 per second out.

From an artificial perspective that would be an increase in power. Quite difficult if not impossible to duplicate by mechanical means without pulleys and ropes or a cantilever off balance center post. Artificial applications of the distribution of energy.

How then is 2 in and 10 out achieved? By the single controlling living biological clock functioning in upside down fashion.

Remember the clocks in the room? All running from a single quartz frequency with each one beginning it's cycle on a different wave of the frequency the results are staggered and blending and therefore seemingly continuous in their output of clicks. Let us use those simultaneous clocks as a logical process and perform a system process using only the clock clicks.

To begin, we establish 100 of the clocks as our input receptors. We remove the second and minute hands from these 100 clocks as they are dedicated to only perform a single task. Then we set them to begin their cycles so that each clock's hour hand clicks twice each second. We do that by assigning each clock a firing pattern of once every 17,500 cycles. Keeping in mind that each click represents a value received by the receptor.

Next, we assign another group of clocks as the first level of memory. These clocks will compare the value of their clicks to the value of the input receptor's clicks at a faster rate. We want these clocks to function at a 30 to 1 ratio to the input clock clicks, so each one is assigned to a firing pattern of every 583 cycles. The minute hand is left on these clocks to click 30 times for each input click or 60 times per second.

Next, we assign another group of clocks as the top level (human) short-term and we use the second hand alone to display this movement. We want this function to operate at a 30 to 1 ratio to the long-term memory's click rate. So we assign each clock to fire it's pattern once every 19 cycles. The result of that computation of values is sent back to the short-term. This 'back action' is what permits humans to be conscious. Aware of a process. It is not the actual awareness of self. That is accomplished by the collection of supported long term memories generated by the feedback loop of the short-term.

The base operating frequency varies from creature to creature with some even and some odd in count. The example shown above is only one such frequency. The value of each 'sample' of memory processed gets smaller and smaller as the memory goes deeper and deeper into the pathway giving rise to the notion of 'past', which gives rise to the notion of 'future' and our concept of time.

This process has built the center post to our brain's teeter-totter. But unlike the conventional artificial teeter-totter the center post of the brain is not a single point in the middle. It is the entire length of the board. Input places a value into the system. Output expels a value out of the system. So the weights of both are not felt by the system.

The input was right side up. The system made it upside down. That means another part of the system is still necessary to 'convert' the values back to right side up.

The hypothalamus, and parts of the thalamus sometimes considered a functionally related collection of parts called the limbic system is thought to be involved particularly with the sense of smell and with certain complex emotional responses, but it also plays a role in regulating basic body functions.

One part of that regulation is in it's conversion of upside down brain functioning values into right side up brain outputs. It operates at 10 cycles per second which means our clocks assigned to this function would fire their pattern every 3500 to 3572 in average cycles. With a steady stream of values entering the limbic system every 583 cycles for long-term process movements and every 19 cycles for short-term process movements the output to muscle groupings and body functions is quite smooth.

Even though the muscles will only function in varying values 10 times per second that is 10 times per second per computational pathway. In between those values is a value sent in opposition of the muscles movement. So that which is sent is a 'don't rest' value.

To Wilson and McNaughton the major difference was that the firing was more rapid, as if the memories were being run on fast-forward. The firing occurred during slow wave sleep. Looking at signals being received from the long-term only brain of a rat, the signals would indeed seem fast.

What did Wilson and McNaughton observe? The long-term memory process of the rat. As fast as the rat's brain can go.

Karni and Sagi disrupted the sleep of volunteers after they had had a training session. Interfering with the subjects' slow-wave sleep had no significant effect. But an equivalent disruption of REM sleep, kept the subjects from improving overnight. Looking at training results being received from the short-term processing brain of a human would indeed interrupt only when the short-term level is active. REM sleep.

What was it that Karni and Sagi observed?

Was it something totally foreign to the foundations made by Wilson and McNaughton?

Not at all. Not foreign, just not the same.

The article then concluded that the experiments lend support to a theory advanced by Jonathan Winson of the Rockefeller University that dreams represent, in effect, "practice sessions" in which animals hone survival skills. Another observational illusion.

Why did Karni and Sagi detect memory consolidation during REM sleep and Wilson and McNaughton only during slow-wave sleep?

When in effect each group studied a different type of memory processing, one involving long-term only (as the rat is not capable of short-term yet acquires REM sleep in a lower process wave) memory comparison that functions at a greater speed than input and one involving short-term memory processing that occurs during REM sleep. Red apples and green apples. Both apples yet from separate trees.

When does all of this start to take place? At the moment the brain is formed. When does all of this cease to take place? At the moment the brain ceases to function. Each, a stage of development of the brain. When the frequency of the biological clock is present, and the wave processed has amplitude a separate life is present.

So what makes the rhythm of the plants differ from us? Plants do not compute their inputs in comparison to previous inputs. They react to their values in comparison to established value norms. Brains compute the inputs and react to the values of them compared to the values of memory. One brain is different from another as one is only able to process based upon previous input. The other is able to process based upon previous comparisons to previous comparisons of previous comparisons.

In between, are the enormous variations of mental abilities and the wonderful levels of the development of life that are the creatures that make this world all that more interesting to ask ....Why?

Chapter Fourteen
Modern Mysticism
The Church of Science Soup

Professor Peter Cochrane is a marvel. He is a man of intellect and stature, of renown and result, of uncommon focus and clever assessment.

Dr. Cochrane is a rare breed in a time of cloned academicians. This was new to me, after pouring over his published materials, as I had never heard of Peter Cochrane until an email brought him to my desk.

A message, carrying the subject of a note I had shot off to the editor of Silicon.Com touting what was then, an article seeking comments from artificial intelligence researchers, actually part of chapter 12 of "The Brain Is A Wonderful Thing" was received from Peter Cochrane with comment that at first glance (and that was as far as the glance got) was not intended for me.

I returned it to the Professor, without reading the balance of the email, and deleted both the original and the 'sent' copy from my computer. So I did not know what it was about or where it was started. Just that it did not address me, even though my address was in the 'to' lines, among others.

Having received his mail, I became aware of his name and searched it. Article after article resulted, mostly from his own web site and Silicon.Com. As I read the material at both locations I was pleased to find an uncommon spirit, a person who observed and questioned, a person who admitted to thinking into the future.

It was not long, after returning the mail that it came back to me in response. That message was directed to me while containing the original email beneath it so then, I read it.

I have intently viewed the major video presentations available at the Professor's personal website and have read the material, even responding with a letter of detail of how the general surface appeared so similar.

The paper, now part of chapter 12 of the book, (as it was intended to be), was titled "The Problem With AI", and perhaps it held some sort of relevance to the other participants of the email that started it all.

During that time I was also working on the actual proposal for The Enticy Institute. After a long time of rewording and concern that the presentation might not be 'up to snuff' it dawned that a business angel had just received far greater detail on the underlying science of charge states, its history and its unlikely outcomes than any publication and who better to judge the worthiness of an institute proposal than a man I had come very quickly to highly admire.

The proposal, secured in a safe directory, was never submitted in that form and an outside opinion would have been wonderful.

Was it written so it could be understood or was it written so I thought it could be understood? What would the 'take' be?

So I asked.

"If you had a Nobel Prize someone might listen."

Well, obviously it was written well enough to cause a relevance to something.

But I still do not know how 'up to snuff' the presentation is.

What I do know is that even the brightest among us, those who lead and are viewed with awe are subjects of human nature.

A brilliant man and a long career of brilliant concepts and a brilliant mind are not the focus of this piece. They are the catalyst. Agree as I do with most of what the Professor writes I would be lax in capacity if I did not use the finest to explain the weakest. Peter Cochrane is a good man. One cannot watch a lengthy keynote address for the World Conference on Technology without gaining some idea of character. Peter Cochrane is very worthy of emulation.

Writing in 'Uncommon Sense": a weekly article for Silicon.Com Peter Cochrane wonders: "No doubt about it, we seem to be moving from a world of concentrated skill and expertise to a world of distributed ignorance." [1]

Cochrane continues by asking, nay expecting: "In this new millennium each of us have to adopt a different mode of operation: to give away, broadcast, make available all our knowledge and information in order to contribute to the broader objectives of our organizations. We should not seek to be powerful and to control; we should seek to be influential and to contribute." [1]

Contribute for the common good. Agree to be a part of the congregation, never disagree to be shunned by its members. Know all you can, so you can help defray those who dare to know more, as anyone who claims to know more is not playing by the rules and should be burned at the stake. It is blasphemy to question established and agreeable theory. Only those who have been previously accepted are currently accepted. Sounds like a form of commune or a church.

Writing in 'Science And Belief', Cochrane says: "Outside the reaches of science we find a multiplicity of belief systems and practices leading to wrong assumptions, decisions and often catastrophes." [2]

Theory is an unproved assumption. It is also a belief. Yet science is ruled by theories, through believers, who know full well their theories lack enough evidence to be law, yet follow them blindly, who know full well, their theories lack foundations, who know full well their theories are based on assumptions. Equal assumptions mean peer agreement. Peer agreement is all peer review is.

Science does not seek truth.

It seeks theories of truth.

"Even if there is only one possible unified theory, it is just a set of rules and equations. What is it that breathes fire into the equations and makes a universe for them to describe?" says Stephen Hawking. [4]

If science sought truth it would settle for no theory. Accept no theory. Base no future science on theory. Find the use of the term "theoretical evidence" to be laughable. Deduce no future deductions on past theory.

It would seek law.

It would seek truth.

If those who fund the church of science wanted results they would only fund those things based in law and the search for law.

If science cannot reach a law they should stop until they do. Basing anything on a foundation of sand is like: "Belief systems, human ignorance, stupidity and vested interests are powerful allies and extremely resilient in all cultures, past and present." [2]

But Cochrane would have one believe the church of science is above all that.

Cochrane claims the reasons to be: "Individuals choose to ignore demand, technology, human nature, practical and theoretical evidence and experience." [2]

Normal humans, not belonging to the robed church of science, know all too well, through experience and practical evidence that theoretical evidence is proof of a theory, not a law, that human nature tends to hold on to what it has held on to before and technology, although enticing and golden glittering as it is, is only the mechanism, not the science.

Outside the reaches of science: "The business world would do well to learn from the kind of verification and peer group review that is central to science." [2]

Really. They do. They just call it a committee. Nothing from a committee ever works. Peer review is like gang membership. You belong on the turf or you do not. Your words match the words known or they are not words worth knowing. Your bigotry is based in peer acceptance at the cost of those who are not peer recognized.

The process of peer review is not double blind testing. That is the method. The process of peer review is human nature based in past human experience. It is fallible.

"And worse, in an IT dominated world that is moving faster, and becoming increasingly non-linear, belief systems are even more damaging," says Cochrane. "Bad decisions now create even worse results in a much shorter time." [2]

So they do.

But as long as one belongs to the church of science the chances of that happening are slim? After all, everything science bases everything on is defined.

Right?

Cochrane disagrees: "Everyday there are words that we use as if we know what they mean. We take action and make decisions as if we had an adequate description, definition, quantification and measure. Here is my short list of the five most important words that we do not understand: life, intelligence, complexity, scalability and value." [3]

If we do not understand 'value': what difference does it make, to define any of the others?

It makes no difference: as evidenced by the words science

uses, relies on and defends that have no meaning.

Entire industries are wrapped around the lack of a definition of intelligence and seek to replicate what no one can bother with defining.

The defenders of the church of science derail and excuse the beliefs of religion without offering so much as a single alternative of law.

Spirit, soul and the rest of the beliefs of life held by varied religions are no more based in observable fact than any theory science has offered to refute them.

The details are argued while the theories that make the belief system of religion are no less theories than the theories of science that dares claim dominance.

Entire research projects costing into the multiple millions of dollars are wrapped around studying the cosmos without a single definition of what it is.

That would require a 'Theory Of Everything'?

NO!

It would require a LAW OF EVERYTHING.

The church of science seeks the theory without regard to any law.

What if the law included science but not in the same theories it has held so close as its belief?

Would science acquiesce?

What if the law included religion but not the same theories it has held so close as its belief?

Would religion acquiesce?

Would truth have merit if it did not fully agree with theory?

Yet the studies continue, as the church of science carries weight among the believers, who trust it to provide knowledge they do not possess, and religion carries weight among the believers, who trust a higher power for the knowledge they do not possess.

Both camps look to the knowledge keepers.

The mystics.

The Priests.

The Shamans.

The Shamans of the church of science's doctrine lead through two means: approval of their level of accepted knowledge (doctrine) and/or the degree to which they adhere to the doctrine while adding to its mysticism.

Dare not dispel the mysticism.

The method is no different in religion.

The means is.

Where the church of science defines causes as smaller sources (looking a lot like marbles, only smaller), religions define cause as the larger source or sources (looking like whatever the era finds desirable).

What's the difference?

Perspective.

Cochrane's personal beliefs lend some insight: "Quantum mechanics looks to be increasingly wrong, and a unified field theory would explain more and make more possible. Antigravity looks to be fundamentally impossible. Matter transport looks feasible. Time travel is a big question as we lack sufficient fundamental understanding. My guess is - it will be possible one day. "[5]

Fundamental understanding would require defining the foundation than as more than sand. Would it not?

"Over 1000 years ago - a god - looked like a reasonable hypothesis. Today it looks like a real long shot! Some central life force - based on fundamentally simple rules - looks to be an increasingly strong possibility!"[5]

Perhaps Cochrane's perception of - a god - is an old man sitting on a thrown?

If there is - a God - would such an entity not be the most simplest form in the Universe or the most complex?

If complex how could simple truth be derived?

Would not a source be the most definitive?

Perhaps it is only a matter of definition?

Perhaps definitions are defined by perspective unless they are true?

On entropy, Cochrane believes it to be: "A celestial ratchet - it goes one way and eventually kills us!" [5]

Does not 'time' do the same thing?

Would 'time' not be the same thing, only perceived by humans to have a 'future' when it is nothing but measurement of entropy, only because the 'future' is a concept and 'time' is the past in our memory?

Depends upon perspective.

Religion is conceptual.

Concepts are believed.

Science is empirical.

Only what is seen is believed.

Could either be infallible?

No.

How could empiricism have been eliminated when science requires observation for belief? Empiricism is otherwise known as quackery.

How could modern science be ruled by logical positivism, when what one sees is directly relational to what one looks for?

Today's science draws logical positivists who require visual stimulus to know. Once they have seen, they have a vision. That vision is reality to them, even if it is not realistic at all.

The process of evaluation, left only to one sense ignores the others and relies on belief to sustain itself. Cochrane says: "Mother Nature is not at all interested in our beliefs - she has her own truths" [6]

If a person is not dependent upon a visual image that needs fulfillment the potential of conceptual interpretation allows evaluation to take place in logic.

Logic is not the science of digital computing, but try to find a logic course in a college anywhere that is not about binary logic.

Binary logic matches the vision.

Words like "life, intelligence, complexity, scalability and value" are comfort to those who use them as intersections to other purposes as they do not need definition. Definition is already in place in the form of a mental image.

They need no definition to represent traffic signs, standing alone without meaning allows all to imply what they wish.

From such implications, grand schemes are concocted which upon evaluation stand on nothing but the word without meaning.

When those schemes find followers, the belief takes on a life, intelligence, complexity, scalability and value of its own.

Explaining the meaning of such words sets definition, where the paths of ignorant progression, dare not permit.

Science is ruled by visual thinking.

In one of the book's chapters I rhetorically asked where all the great aural thinking scientists have gone.

Since then I have surmised them to be either in retirement, or industry.

Academics are mostly visual as **tenure needs no**

performance and therefore there are **no consequences for promises made in fields where definitions of the fundamentals are not required**.

The acceptance of theory as the goal of science is tantamount to any military declaring victory over an opponent simply for having designed a battle plan.

Believing in theory is as if that battle plan was the basis for each subsequent battle plan even though it was never proven true.

That is not to say that all scientists are visual thinkers.

And it is not to say that visual thinking is bad.

It is not.

It is just not capable of creativity beyond what has already been created.

It makes adaptations.

It makes interpretations.

It makes improvements.

Coupled with an advanced aural process in long-term memory it results in the ability to explain a concept by a visual image that is already understood.

It makes for redundancy in agreements if the knowledge of a visual image has become widespread.

The image becomes the point and not the concept that used the image only as an example.

The more knowledge acquired, the more the image is engrained; the more defense of it is necessary to remain coherent, balanced and comfortable.

When 'comfortable' becomes the norm all new things are viewed with the tiny focus of the comfort of what is already known.

Science has become what it detests.

It has grown in knowledge but it has grown on that

knowledge. It has not created new knowledge is quite a long time.

Relying on theories to build new explanations means: "If your only tool is a hammer, every problem looks like a nail" [6]

As much as mathematics is logic in order it is indeed: "... fundamentally a visualization tool" [6]

In "Intelligent Machines" Peter Cochrane says: "I think what we have is rapidly becoming boring. For sure, it is magical in its performance and abilities but nothing much fundamental has changed for the last five years. Yes, our devices have become smaller and displays are bigger, and interfaces have become a little friendlier but the reality is that the PC, PDA, mobile phone and laptop are pretty much the same devices as those we had five years ago. The question is: for how long will this continue?" [7]

"By 2020 I expect all our devices to be making intelligent decisions about steering messages across a room or through a building instead of the dumb routings of today. I expect location-based activities and state to be subsumed into my devices so they can make sensible decisions about my travel, work and communication. I need far less overload and far more effectiveness and only intelligent machines can give me that. I need technology to augment my existence, to detect when I am tired, hungry or on a roll, so I can be automatically steered towards the right activities and decisions." [7]

Would it be wise to build a machine that thinks and is self-aware?

Yes, to make mankind 'aware' of the process of 'why' and 'how'; to employ the same system in the laboratory to make healing part of the system; to prove that mankind has the same ability in mind.

Both 'churches' science and religion, have disagreement on that.

Religion fears intelligent machines for their potential to take over intelligent mankind.

Not if mankind knows how intelligence works and therefore how everything works.

Science is trying, but has no definition of intelligence to base it on and is continually trying not to reach a goal by refusing to admit that today is yesterday's future.

Everything worth reaching is still considered 'in the future'.

There is a common method for science to act and a not so

common method.

Peer review of any uncommon science is like making soup.

It is the method that has kept major scientific breakthroughs a long process and relegated science to the study of its own theories instead of the study of the laws they seek:

Science Soup:
Ingredients

1 cup theory
½ tsp knowledge
4 large cups speculation
½ tsp contemplation

Place speculation and theory into large cooking pot: place on shelf for 10 years.

Resurrect pot and bring to a boil. Add knowledge and let simmer another 10 years.

Stir only once.

After simmering, add ½ tsp contemplation, again stirring once.

Let stand until it proves itself.

In the mean time, refer to the pot of theory as fact and defend it at all costs while making more pots based on the most accepted pot.

Only break the seal and serve the soup when empirical evidence claims it has already happened.

"A long habit of not thinking a thing wrong gives it a

superficial appearance of being right." Thomas Paine [7]

"Nothing is more difficult than to introduce a new order. Because the innovator has for enemies all those who have done well under the old conditions and lukewarm defenders in those who may do well under the new" - Nicolai Machiavelli, (1513)

"Any Sufficiently advanced technology is indistinguishable from magic" (Arthur C Clarke) [7]

"We know very little, but are capable of a great deal." Peter Cochrane [7]

When asked "Is there something we have missed - one crucial development - that will be obvious in 10 or 20 years time but which now is considered irrelevant or marginal?" Peter Cochrane answered, "Artificial Life". [8]

One of those words that will remain undefined until sometime in the future?

No.

Just one of those words that will remain undefined, as to define it, would bring the future, and end the past.

Belief requires mysticism, ineffable terms and comfort with not knowing.

I do greatly admire the work of the Professor, but I am a bit confused:

I've never considered a business proposal to require a prize of any sort.

Bibliography

Chapter 2

[1] **How the brain processes emotions** Monday, 13 January, 2003, 00:08 GMT BBC NEWS http://news.bbc.co.uk/1/hi/health/2635269.stm
[2] **Angry outbursts linked to brain dysfunction** 22:00 27 May 02 New Scientist. http://www.newscientist.com/news/news.jsp?id=ns99992331
[3] **Brain expression response linked to personality** 19:00 20 June 02 New Scientist http://www.newscientist.com/news/news.jsp?id=ns99992439
[4] **Brain scans can reveal liars** 10:50 12 November 01 New Scientist http://www.newscientist.com/news/news.jsp?id=ns99991543
[5] **Mind theory** 13:50 29 March 01 New Scientist http://www.newscientist.com/news/news.jsp?id=ns9999567
[6] **Brain's 'cheat detector' is revealed** 22:00 12 August 02 New Scientist http://www.newscientist.com/news/news.jsp?id=ns99992663
[7]Chewing gum improves memory14:30 13 March 02 New Scientist http://www.newscientist.com/news/news.jsp?id=ns99992039
[8] **Dream machine** 10:10 13 October 00 New Scientist http://www.newscientist.com/news/news.jsp?id=ns999970
[9] http://faculty.washington.edu/chudler/ap.html

Chapter 3

[1] http://www.stanford.edu/class/history13/earlysciencelab/body/brainpages/brain.html
[2] http://www.rader.wramc.amedd.army.mil/psych/brain%20functions.htm
[3] http://www.dianetics.org/en_US/articles/feelings/stored.html
[4] http://epswww.unm.edu/facstaff/zsharp/106/lecture%202%20steno.htm
[5] http://www.wikipedia.org/w/wiki.phtml?search=Schizophrenia&go=Go
[6] http://www.wikipedia.org/wiki/Causes_of_psychiatric_disorder
[7] http://www.wikipedia.org/w/wiki.phtml?search=mind&go=Go
[8] http://www.wikipedia.org/wiki/Dream
[9] http://www.wikipedia.org/wiki/Nightmare
[10] http://www.wikipedia.org/wiki/Brain_event
[11] http://www.wikipedia.org/wiki/Thomas_Nagel
[12] http://www.wikipedia.org/w/wiki.phtml?search=consciousness&go=Go
[13] http://www.wikipedia.org/wiki/Cognitive_science
[14] http://faculty.washington.edu/chudler/tenper.html
[15] http://www.wikipedia.org/wiki/Cognitive_bias
[16] http://www.memorylossonline.com/memorytip.htm
[17] http://www.sciencenet.org.uk/database/Social/Original/s00007d.html
[18] http://science.howstuffworks.com/brain1.htm

[19] http://www.wikipedia.org/wiki/Complex_system
[20] http://www.dai.ed.ac.uk/homes/cam/IAS_docs/BBAss/B_B_Notes/node3.html
[21] http://xtronics.com/memory/how_memory-works.htm
[22] http://www.class.uidaho.edu/mickelsen/texts/marxpap.htm
[23] http://faculty.washington.edu/chudler/sleep.html
[24] http://www.sciencenet.org.uk/database/Biology/Brain/b00387c.html
[25] http://www.ecs.soton.ac.uk/\~harnad/Papers/Harnad/harnad98.explaining.consciousness.html
[26] http://www.culture.com.au/brain_proj/CONTENT/CHAL_01.HTM
[27] http://www.papert.org/articles/Papertonpiaget.html
[28] http://www.buzan.com.au/book_head_first.htm
[29] http://www.plumebleue.ch/pages/MozartEffectEng.htm
[30] http://www.wikipedia.org/wiki/Alfred_Binet
[31] http://www.hon.ch/News/HSN/512674.html
[32] http://www.nature.com/nsu/010315/010315-10.html
[33] http://www.time.com/time/europe/webonly/londoneye/2000/04/grandintvu.html
[34] http://www.noldus.com/events/mb2000/program/abstracts/savoy.html
[35] http://www.encyclopedia.com/html/G/Gestalt.asp
[36] http://www.m-w.com/cgi-bin/dictionary
[37] http://bellarmine.lmu.edu/faculty/mmills_fp/Evolpsyc/spring03/pinker-gender.htm
[38] http://www.wikipedia.org/wiki/Obsessive-compulsive_disorder
[39] http://www.ohiou.edu/\~ridges/history.html
[40] http://www.sspnet.org/public/articles/index.cfm?Cat=4
[41] http://www.ai.mit.edu/people/brooks/books-movies.shtml#cambrian
[42] http://www.ai.mit.edu/people/brooks/papers/CMAA-group.pdf
[43] http://www.ai.mit.edu/people/brooks/papers/representation.pdf
[44] http://www.ai.mit.edu/people/brooks/papers/AIM-1293.pdf
[45] http://www.ai.mit.edu/people/brooks/papers/group-AAAI-98.pdf
[46] http://www.mirecc.org/science-articles/how-neurons-communicate-archive.shtml
[47] http://www.albany.net/\~tjc/neuron.html
[48] http://apu.sfn.org/content/Publications/BrainBriefings/mind.body.html
[49] http://www.marxists.org/reference/subject/philosophy/works/ge/brentano.htm
[50] http://www.wikipedia.org/wiki/Intelligence
[51] http://www.wikipedia.org/wiki/Rene_Descartes
[52] http://www.wikipedia.org/wiki/Cartesian_coordinate_system
[53] Synchronous Firing Of Retina and Neutronics Flat Retina Correcting Camera, Hempfling http://www.enticypress.com
[54] Fuzzy Logic and CORE® Hempfling http://www.enticypress.com
[55] http://www.sffworld.com/authors/m/moy_chris/articles/futureofail.html
[56] http://www.marketscreen.com/help/atoz/default.asp?Num=46

Chapter 4

[1]
http://www.interdys.org/servlet/compose?section_id=5&page_id=9
5
[2]
http://www.interdys.org/servlet/compose?section_id=5&page_id=5
0

Chapter 6

[1] http://www.robin-williams.com/bio.htm provided by Celebrity-Websites.Com.
[2] http://colin.mochrie.com/bio.html
[3] http://www.billhicks.com/bio.html

Chapter 8

[1] Wikipedia http://www.wikipedia.org/w/wiki.phtml?search=rectifier&go=Go
[2] m-w.com Meriam-Webster Online.
[3]   Drug   Effects   on   the   Synapse,
http://psych.athabascau.ca/html/Psych402/Biotutorials/13/part1.html
[4]   How   Antidepressants   Work,
http://www.4therapy.com/consumer/conditions/item.php?uniqueid=23
[5] How Stuff Works: Diodes http://electronics.howstuffworks.com/diode3.htm
[6] Reliability of Spike Timing in Neocortical Neurons. Zachary F. Mainen and Terrence J. Sejnowski, Howard Hughes Medical Institute, Salk Institute for Biological Studies, Departments of Neuroscience and Biology University of California, San Diego.
[7] Merriam-Webster Dictionary.
[8] The Biological Clock, Lee Kent Hempfling, Enticypress 1996
[9] Biological vs. Mechanical Clock Lee Kent Hempfling, Enticypress 1996
[10] The Human Visual Process Resulting In Facial Recognition Using Correlational Opponent Ratio Enhanced Processing, Lee Kent Hempfling, Enticypress 1996

Chapter 9

[1]   http://www-gap.dcs.st-and.ac.uk/\~history/Mathematicians/Thomson.html
[2] http://rinkworks.com/said/predictions.shtml

Chapter 10

[1]   Sleep   and   Brain   Chemistry

The Brain Is A Wonderful Thing
http://www.sciencecentral.com/articles/view.php3?language=english&type=article&article_id=218392013
[2] Short-term dyslexia treatment strengthens key brain regions
http://www.eurekalert.org/pub_releases/2003-07/aaon-sdt071503.php
[3] A Computer in Your Head? by Dr. Eric H. Chudler Originally published in ODYSSEY magazine, 10:6-7, 2001 (March), Cobblestone Publishing Co. http://faculty.washington.edu/chudler/computer.html
[4] Insight Into The Way Pain Is Regulated In The Brain Could Lead To New Target For Therapy
http://www.sciencedaily.com/releases/2003/07/030723084907.htm
[5] Study Links Genes to Depression, Stress
http://paktribune.com/news/index.php?id=32856
[6] http://64.176.52.217/enticypress/genes/index.cgi
[7] http://www.wired.com/wired/archive/11.08/view.html?pg=3

Chapter 11

[1] Merriam Webster http://www.m-w.com

Chapter 13

Ahlgren, Andrew, and Halberg, Franz. Cycles of Nature (Natural Science Teachers, 1990).
Angier, Natalie. Modern Life Suppresses an Ancient Body Rhythm. (New York Times 1995)
Associated Press, Advance Reported On Internal Clock Gene. (Boston Globe 1994)
Foreman Judy, Globe Staff. Boston Team Pinpoints Man's Biological Clock. (Boston Globe 1988)
Glass, Leon, and Mackey, M.C. From Clocks to Chaos (Princeton Univ. Press,1988).
Jespersen,James . Watch And Clock. (Comptons Encyclopedia 1995)
Lerner, Eric J. Early History of Timekeeping. (Comptons Encyclopedia 1995)
Leutwyler, Kristin. Depression's Double Standard. (Scientific American Library 1995)
Waterhouse, J.M., and others. Your Body Clock (Oxford Univ. Press, 1990).
Winfree, A.T. The Timing of Biological Clocks (Scientific American Library,1993).

Chapter 14

[1] http://www.silicon.com/opinion/164-500001/1/5397.html
[2] http://www.silicon.com/opinion/500020/1/1036858.html
[3] http://www.silicon.com/opinion/500007/1/1034472.html
[4] On The States of Energy, Gravity & The Exponential Universe
[5] http://www.cochrane.org.uk/opinion/views/future.htm
[6] http://www.cochrane.org.uk/inside/quotes.htm

The Brain Is A Wonderful Thing

[7]   http://www.silicon.com/opinion/500015-500010/1/3805.html
[8] http://www.cochrane.org.uk/opinion/qanda.htm

Glossary:

Chapter 2

Brain: the portion of the vertebrate central nervous system that constitutes the organ of thought and neural coordination, includes all the higher nervous centers receiving stimuli from the sense organs and interpreting and correlating them to formulate the motor impulses, is made up of neurons and supporting and nutritive structures, is enclosed within the skull, and is continuous with the spinal cord through the foramen magnum

Recall: 1: to have a recollection or remembrance and 2: physical acquisition of previous memory

Dynamic: 1: of or relating to physical force or energy, 2: marked by usually continuous and productive activity or change, 3: requiring periodic refreshment of charge in order to retain data

System: 1: a regularly interacting or interdependent group of items forming a unified whole, 2: a group of interacting bodies under the influence of related forces, 3: a group of biological devices or artificial objects or an organization forming a network especially for distributing something or serving a common purpose

Know: 1: To be aware of, 2: to recognize as being the same as something previously known

Aware: To know that one knows (does not mean 'I am aware of that' meaning a recognition of a previous input, or simple recall).

Data: Information in quantifiable form

Evolution: 1: a process of change in a certain direction, 2: a process of continuous change from a lower, simpler, or worse to a higher, more complex, or better state

Amplitude: the maximum departure of the value of a wave from the average or base value

Long-Term: memory both storing and processing involving a relatively long period of time that reduces in amplitude by the depth of processing

Short-Term: memory both storing and processing involving a relatively short period of time that reduces in amplitude by the depth of processing

Human Short-Term: memory both storing and processing involving a relatively short period of time that reduces in amplitude by the depth of process that loops its final computational output to the beginning of the input to that level of memory causing growth of similars to reach the point of near half the input value to short-term from long-term and builds awareness through the continuation of the same or nearly same amplitudes.

Pathway: a line of communication through processing over interconnecting neurons representing a specific process. There are thousands of such pathways for each type of input receptor.

Regeneration: an act or the process of regenerating

Cognition: awareness of the existence of a process.

Exponential: characterized by or being an extremely rapid increase, a compounded process

Exponential processing: the process expands with each process

Non-Zero: A quantity which does not equal zero is said to be nonzero. A real nonzero number must be either positive or negative, and a complex non-zero number can have either real or imaginary part non-zero.

On The States Of Charge, Gravity and The Observable Exponential Universe: Forthcoming paper on gravity, charge state and unification.

Good: The maximum extreme of positive potential. Reality to the subject is dependent upon previous supported long-term memory which determines what a positive potential might be based on what the most consistent positive potential has been.

Bad: The maximum extreme of negative potential. Reality to the subject is dependent upon previous supported long-term memory which determines what a negative potential might be based on what the most consistent negative potential has been.

Glossary Chapter Fourteen:

No matter how hard science tries, words mean what words mean:

From Merriam-Webster:

Church: a body or organization of religious believers:

Common Sense: 1: the unreflective opinions of ordinary people 2: sound and prudent but often unsophisticated judgment

Cosmology: 1 a: a branch of metaphysics that deals with the nature of the universe b: a theory or doctrine describing the natural order of the universe 2: a branch of astronomy that deals with the origin, structure, and space-time relationships of the universe; also: a theory dealing with these matters

Doctrine: a: something that is taught b: a principle or position or the body of principles in a branch of knowledge or system of belief: c: a principle of law established through past decisions

Empirical: originating in or based on observation or experience

Empiricism: a former school of medical practice founded on experience without the aid of science or theory

Entropy: the degradation of the matter and energy in the universe to an ultimate state of inert uniformity

Ineffable: incapable of being expressed in words

Knowledge: the fact or condition of knowing something with familiarity gained through experience 'or association

Logical Positivism: a 20th century philosophical movement that holds characteristically that all meaningful statements are either analytic or conclusively verifiable or at least confirmable by observation and experiment and that metaphysical theories are therefore strictly meaningless -- called also logical empiricism

Priest: one authorized to perform the sacred rites of a religion

Religious: relating to or manifesting faithful devotion to an acknowledged ultimate reality or deity

Science: knowledge or a system of knowledge covering general truths or the operation of general laws especially as obtained and tested through scientific method

Scientific Method: principles and procedures for the systematic pursuit of knowledge involving the recognition and formulation of a problem, the collection of data through observation and experiment, and the formulation and testing of hypotheses

Shaman: a priest or priestess who uses magic for the purpose of curing the sick, divining the hidden, and controlling events

True: being in accordance with the actual state of affairs; being that which is the case rather than what is manifest or assumed

Mysticism: vague speculation: a belief without sound basis b: a theory postulating the possibility of direct and intuitive acquisition of ineffable knowledge or power

Theory: a belief, policy, or procedure proposed or followed as the basis of action; a plausible or scientifically acceptable general principle or body of principles offered to explain phenomena; a : a hypothesis assumed for the sake of argument or investigation b: an unproved assumption.

INDEX:

.